MAKE UP PATTERN DESIGN BOOK

메디시인

필자소개

김 민 경
장안대학교 교수

곽 지 은
경북과학대학교 교수

김 선 희
수원과학대학교 교수

안 은 주
유한대학교 교수

이 주 미
인천재능대학교 교수

홍 수 경
동남보건대학교 교수

국가자격증
메이크업 패턴 디자인북 개정판

2018년 11월 20일 초판 발쇄
2022년 2월 10일 개정판 발행

지은이 • 김민경 · 곽지은 · 김선희
안은주 · 이주미 · 홍수경
발행인 • 주병오 · 주민기
발행처 • 메디시인
주 소 • 경기도 파주시 회동길 209
파주출판문화정보산업단지
영업부 • 031-955-7566 · 7577
편집부 • 031-955-7731
F A X • 031-955-7730
홈페이지 • www.ji-gu.co.kr
전자우편 • jigupub@hanmail.net
등록번호 • 2005년 2월 16일 제 406-2005-000045호

ISBN 979-11-86739-70-9 가격 : 25,000원

머리말

메이크업 패턴 디자인 북은 메이크업 실기 국가자격증이 시행됨에 따라 자격증 취득에 요구하는 12가지 패턴의 일러스트를 넣어, 메이크업 기능사 자격증을 준비하는 사람은 물론 메이크업에 입문하는 학생 및 전문가 등의 사람들이 활용할 수 있도록 구성하였다.

또한 이 패턴북은 기초 뷰티 메이크업부터 패션 & 판타지 메이크업, 페이스 & 바디 페인팅 뿐만 아니라 헤어 연출 방법을 포함하여 표현방법을 향상하는데 도움이 되고자 하였다.

메이크업 실기에 앞서 디자인 능력을 키우는데 도움이 되고자 하였으며 메이크업 기초부터 심화과정에 이르러 다양한 연습을 할 수 있도록 제안하였다.

따라서 패턴북이 제시하는 여러가지 타입의 디자인을 꾸준히 연습한다면 메이크업 작품 완성도를 높이는 것은 물론 메이크업 국가자격증 취득과 더불어 직무 수행에도 도움이 될 것으로 확신한다.

이 책은 메이크업 전공을 희망하는 학생들과 더불어 현장에서 종사하는 메이크업 아트스트에게 미력하나마 작은 도움이 되길 희망한다.

이 책이 출판될 수 있도록 많은 도움을 주신 (주)지구문화/메디시언의 주민기 이사님을 비롯한 편집에 도움을 주신 임직원 여러분과 김태은 선생에게 감사드린다.

임인년 삼월에

저자 일동

목 차

Chapter 1

뷰티 일러스트 탐구

• 재료

드로잉에 사용되는 재료는 여러가지가 있어서 부드러움에서부터 날카로운 느낌에 이르기까지 다양하게 표현할 수 있다. 주로 연필, 색연필, 파스텔, 목탄연필, 아이 쉐도우 등으로 부드럽게 표현하며, 그 외 수채화 물감, 아크릴 물감, 에어 브러쉬 등을 이용하여 다양하게 연출한다. 여러 가지의 재료들 중 한가지만으로 표현할 수 있으나 각각의 재료를 혼용하여 사용하기도 한다.

• 연필

흑연 연필, 샤프, H~6B에 이르기까지 심의 경도에 따라서 부드러움과 강한 느낌의 표현을 연출할 수 있다. 일반적으로 B로 표기된 심은 부드럽고 진하게 표현할 때 사용하며 H로 표기된 심은 단단하고 거친 느낌으로 표현할 때 사용한다.

9B	8B	7B	6B	5B	4B	3B	2B	B	HB	H	2H	3H	4H	5H	6H	7H	8H	9H

부드러워짐 ◀- - - - - - - - - -▶ 단단해짐

연필심의 경도

① 색연필 : 다양한 색의 종류가 있어 다채롭게 표현이 가능하며 여러색을 겹치거나 섞어서 사용하기도 한다. 색연필의 심을 뾰족하게 깎아 날카롭게 표현할 수 있는 반면 뭉툭하게 깎아서 부드럽게 선을 표현하여 면처리를 할 수 있다.

색연필에 물을 섞어서 수채화 느낌으로도 표현 가능하며 밝은 색부터 어두운 색에 이르기까지 명암을 표현하기 용이하다. 색연필 사용은 밝은 색부터 사용하여 어두운 색 순으로 진항하면 깨끗한 표현을 연출 할 수 있다.

② 아이 쉐도우 : 부드러운 면을 표현하기에 용이하며 브러쉬를 이용하여 여러 번 겹쳐서 색을 표현하면 부드럽게 연출할 수 있다. 가루 날림이 심하므로 가루를 털어가면서 종이에 바르도록 한다.

목탄연필 : 부드러운 특성이 있어 많이 사용되나 사용 후 정착액을 뿌려줘야 한다. 선과 명암의 표현에 탁월하며 단단한 H계열 제품과 부드러운 B계열 제품으로 나뉜다.

③ 지우개 : 하일라이트를 표현할 수 있으며 선적으로나 면적으로 표현 가능하다. 모서리를 잘라서 뾰족하게 다듬은 후 날카로운 부분을 묘사할때 사용하면 용이하다.

④ 카펜터 펜슬 : 2B, 4B, 6B 총 세가지가 있으며 연필 끝이 평편한 직사각형이기 때문에 설계도나 문자도안에 용이하다. 연필심을 의도하는 모양으로 깎아서 사용하면 원하는 느낌으로 선을 연출할 수 있다.

⑤ 수채화 물감 : 물감에 물의 양을 조절한 후 농담의 효과를 나타내어 효과를 다양하게 활용할 수 있는 재료가 수채화 물감이다. 투명한 느낌 또는 불투명한 느낌을 이용하여 다양한 이미지로 표현할 수 있다. 기법의 다양성을 활용하여 번짐, 흘러내림 등으로 회화적 느낌까지 이끌어 낼 수 있다.

Chapter 2

인체의 기본 형태와 구조

물체의 기본 형태는 구, 원추, 원기둥으로 이루어져 있다. 인체도 역시 각 부위를 하나씩 분류해보면 4가지 구조형태로 되어 있다. 인체는 기본적으로 원통형 구조에서부터 시작되고 신체 각 부위 하나하나 나누어 볼 경우 더욱 그렇다.

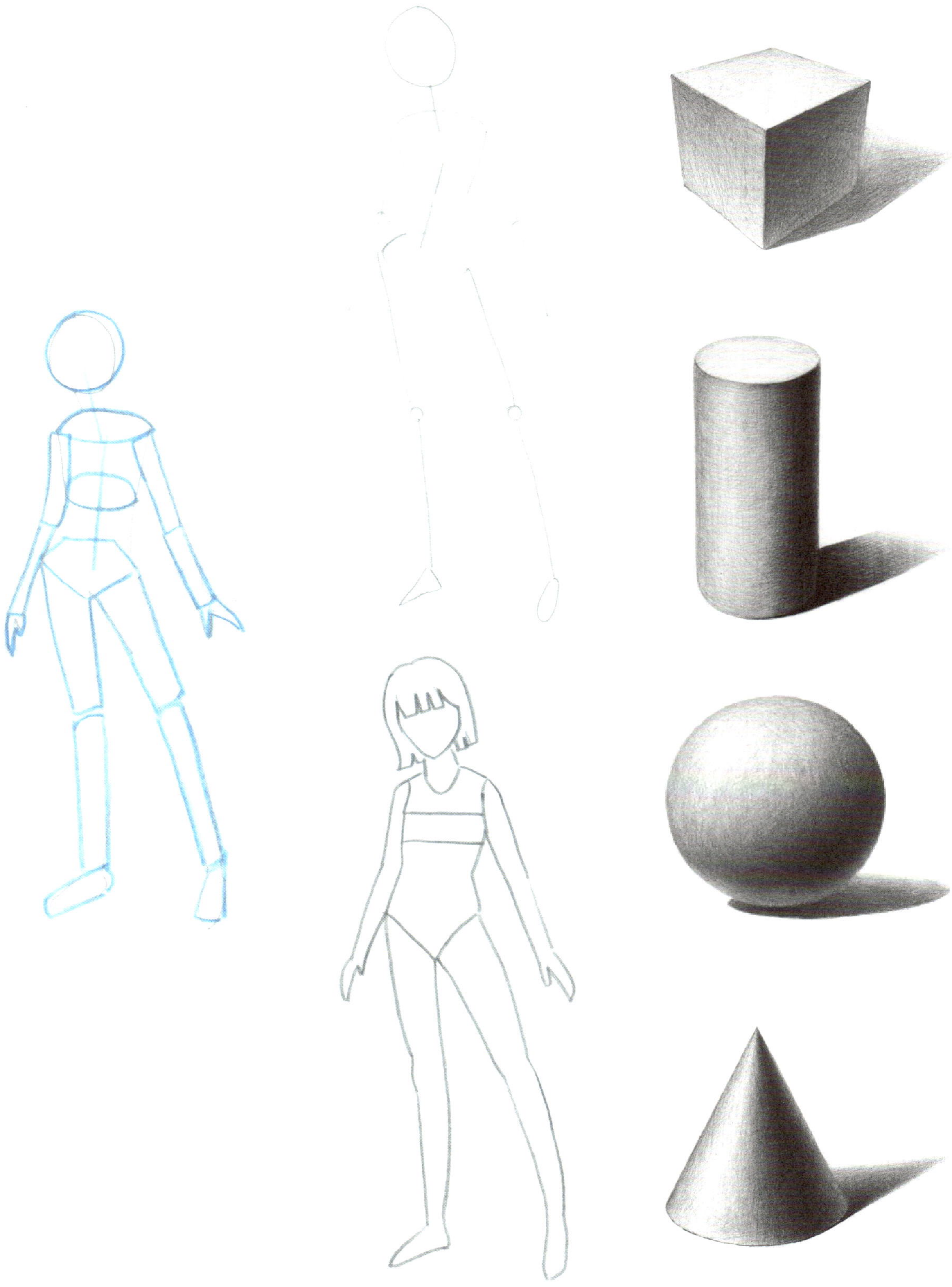

Chapter 3

입술 표현

입술은 중심선을 기준으로 상하좌우의 크기와 위치를 정하여 대칭을 이루게 한다.

윗입술과 아랫입술의 비율은 1:1.5정도로 그리면 도톰한 입술의 형태로 균형을 잡을 수 있다.

아래에서 위로 입술의 중앙정도까지 밝아지다 다시 입술의 안쪽으로 어두워지도록 하여 양감을 표현한다.

Chapter 4

헤어 표현

머리카락의 입체감을 표현하기 위하여 밝은 부분과 어두운 부분의 표현이 중요하다. 머리카락은 두피에서 자라나오므로 반드시 두피에서 헤어를 구부린 형태로 그려야 한다. 그리고 머리카락의 모양을 파악한 후 몇 개의 덩어리감으로 표현하여야 한다.

머리카락의 볼륨은 밝고 어두운 명암으로 처리하며 외형이 둥글게 되도록 그려야 한다. 빛의 방향에 따라 어두워졌다 다시 점점 밝아지는 유형으로 반복되며 양감을 표현한다.

빛이 들어간 부분은 가장 밝게, 머리카락이 만나는 부위는 어둡고 강하게 표현한다.

나선형의 컬 모양은 꺾어지는 부분을 어둡고 강하게 표현하며 특히 머리가 난 방향을 보고 하이라이트로 방향성을 나타낸다.

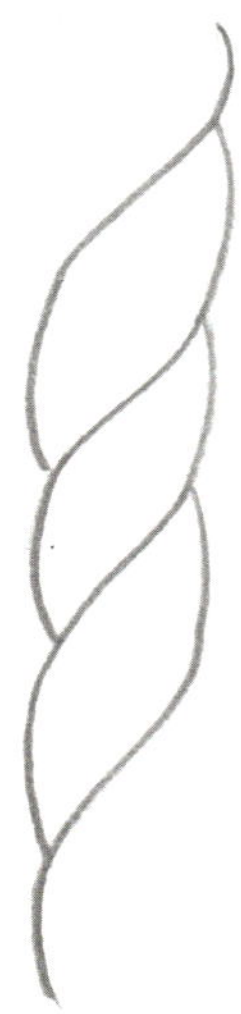

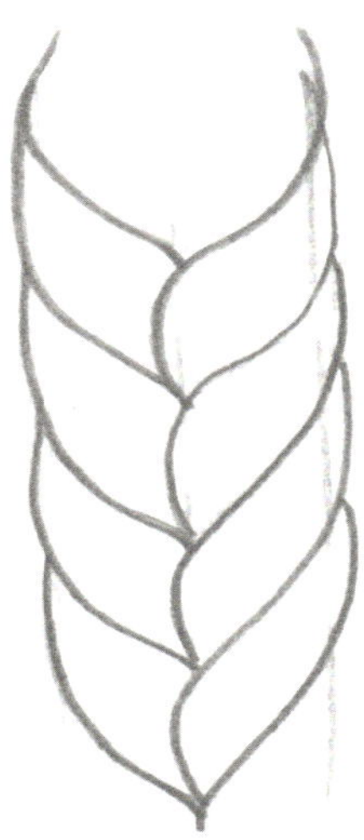

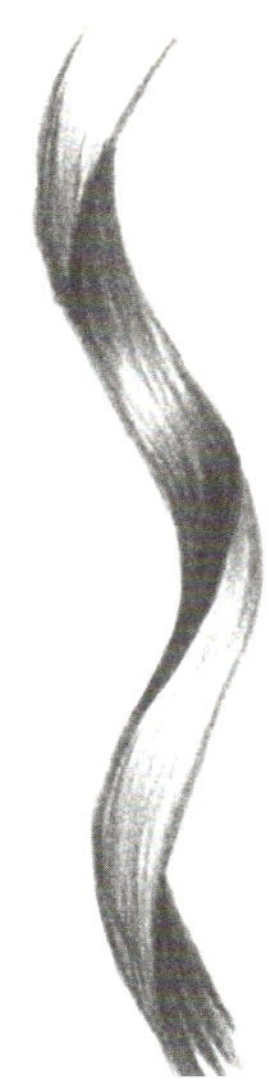

Chapter 5

선과 명암의 그라데이션

명도란 색의 밝고 어두운 정도를 말하며 명도(value)단계는 0~10까지 11단계로 나눈다. 밝은 색은 고명도, 어두운 색은 저명도라고 하며 흰색이 10으로 가장 명도가 높다. 순수한 검정색과 흰색은 명도가 없다. 그러므로 검정색 명암단계는 0.5, 흰색은 9.5이다. 명암은 일정한 방향의 빛이 닿으면 밝은 곳과 어두운 곳으로 나뉘게 된다. 물체는 빛에 의하여 받는 각도에 따라 일정한 질서와 톤을 형성하며 평면에서 입체와 같은 조형을 형성한다.

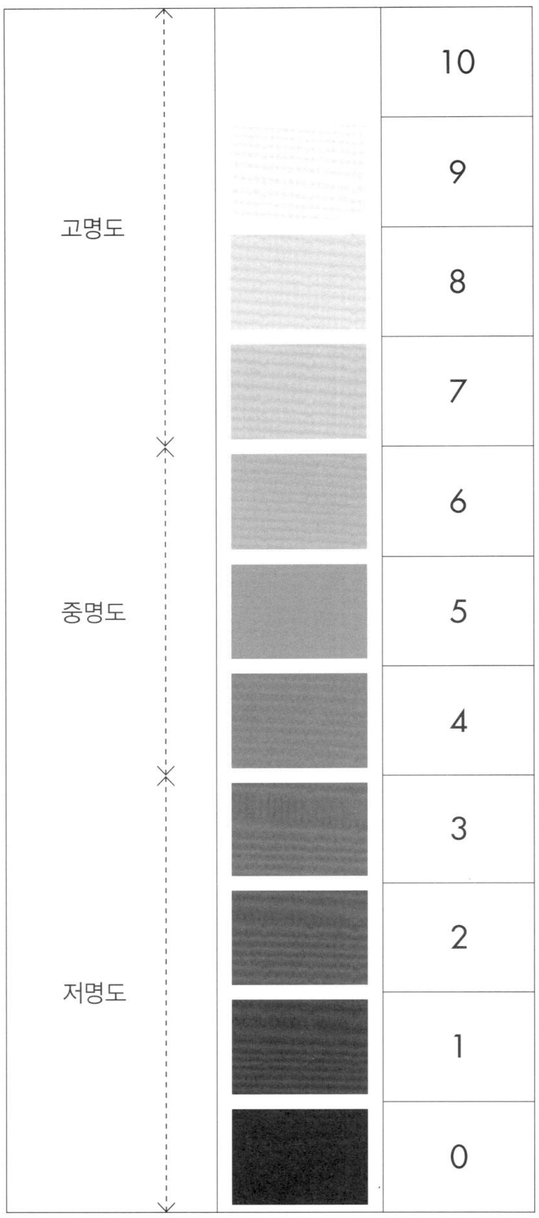

선과 명암의 그라데이션 연습

Chapter 6

얼굴의 구조

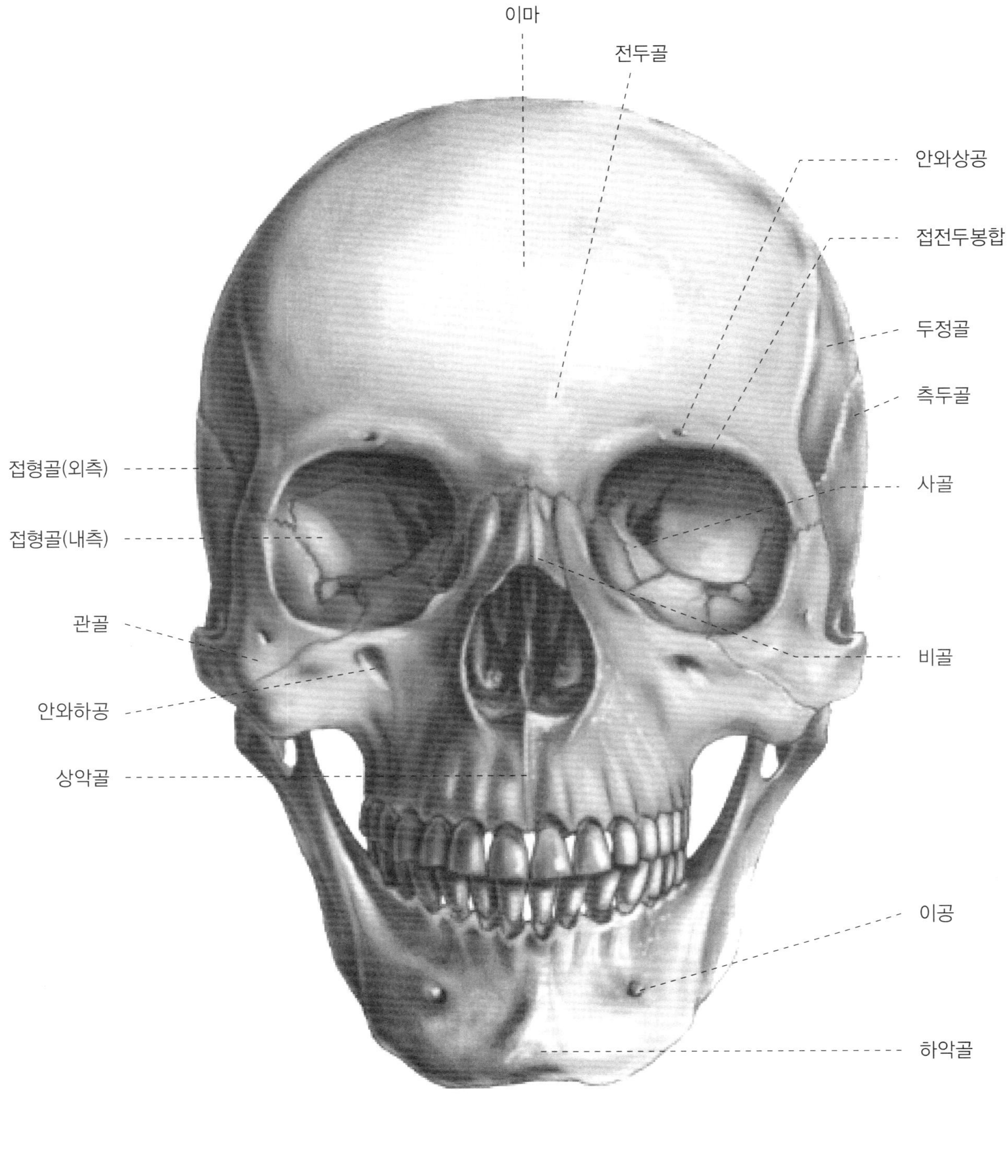

Chapter 7

얼굴의 균형

가로 균형 : 헤어라인에서 눈썹, 눈썹에서 콧망울, 콧망울에서 턱선까지 3등분한다.
세로 균형 : 헤어라인에서 눈꼬리, 눈꼬리에 눈앞머리까지, 눈앞머리에서 눈앞머리까지, 눈앞머리에서 눈꼬리까지, 눈꼬리에서 헤어라인까지 5등분한다.

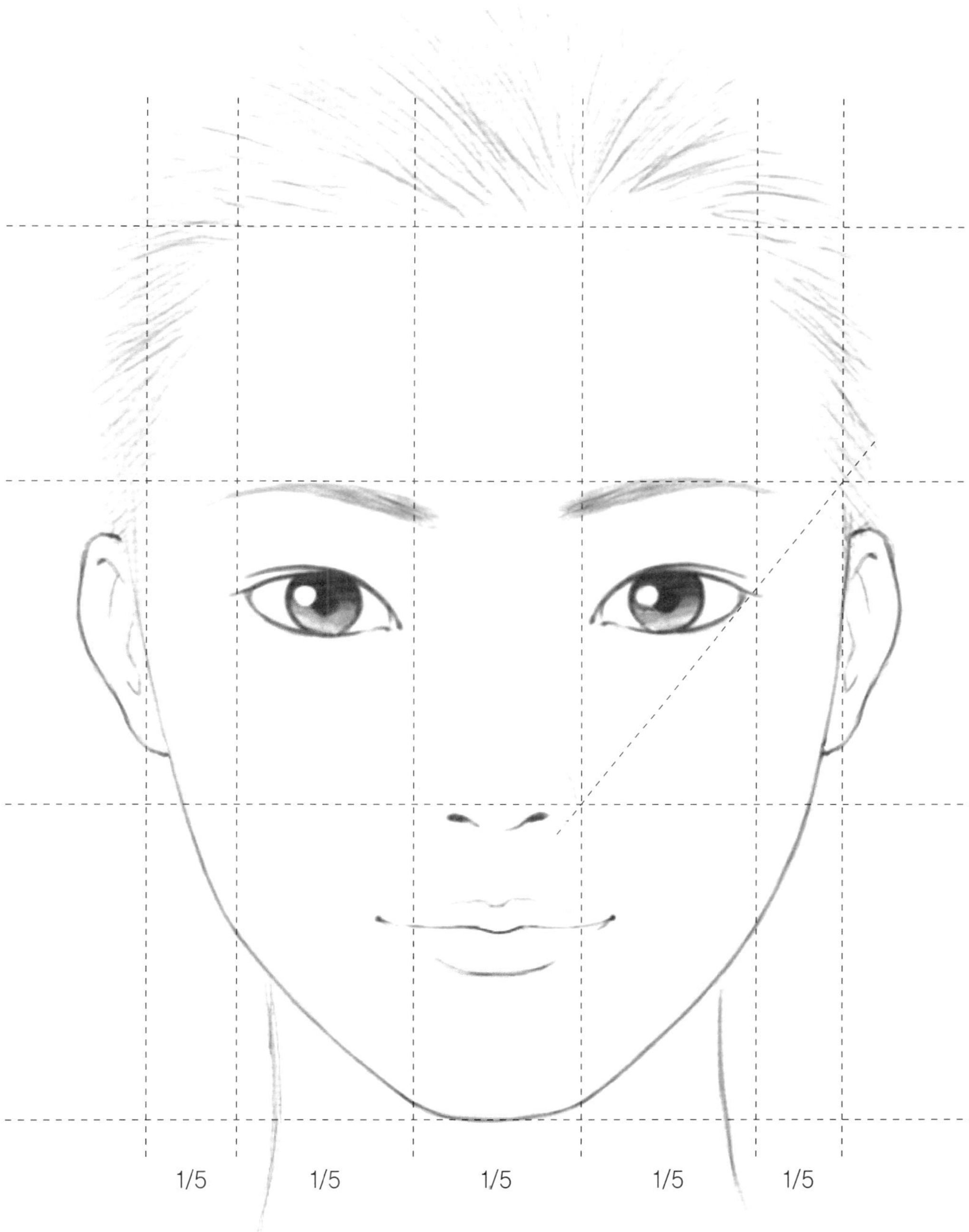

Chapter 8

얼굴의 형태

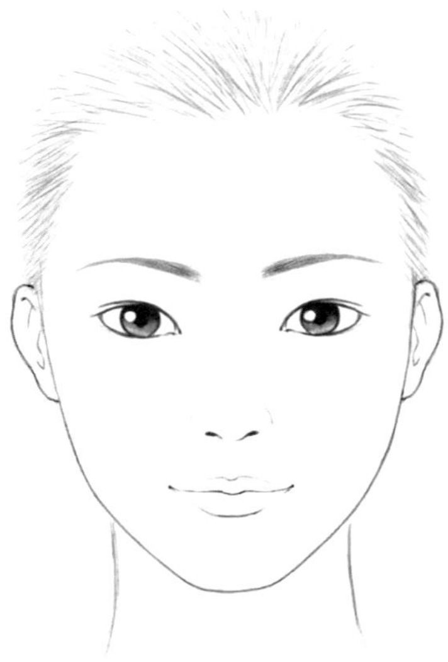

기본형(타원형)

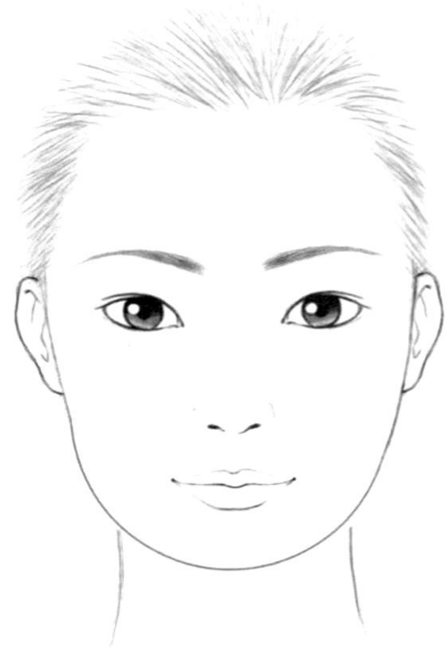

둥근형

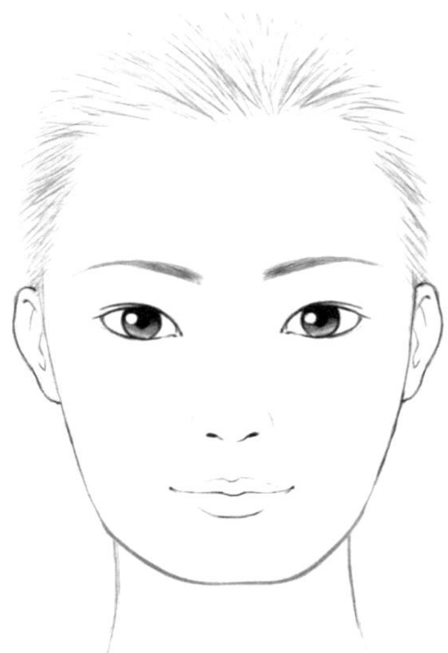

사각형 얼굴

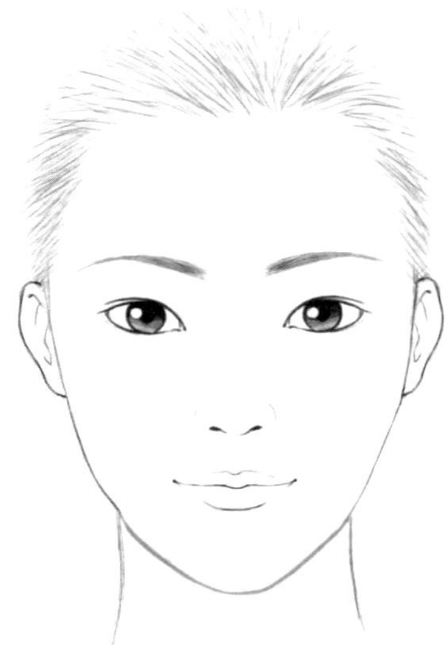

역삼각형 얼굴

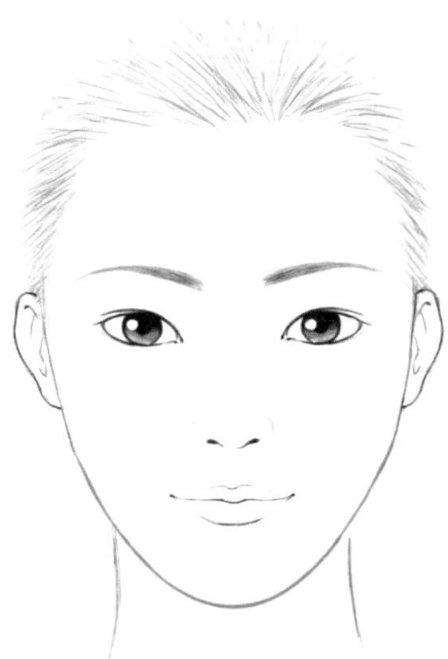

다이아몬드형 얼굴

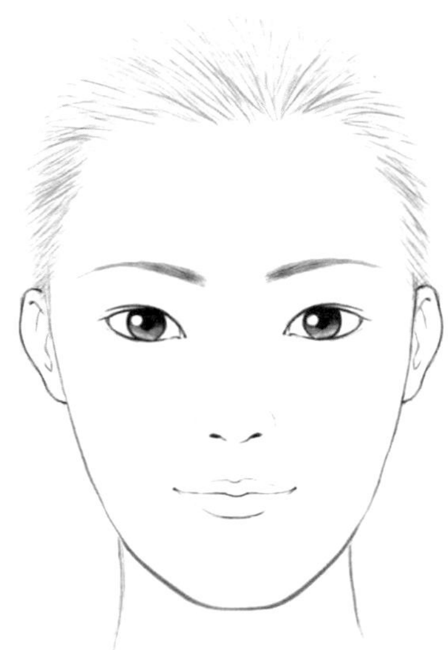

긴얼굴

Chapter 9

Eye-Brow Style

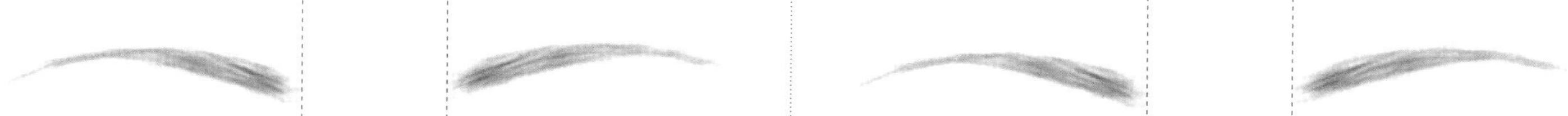

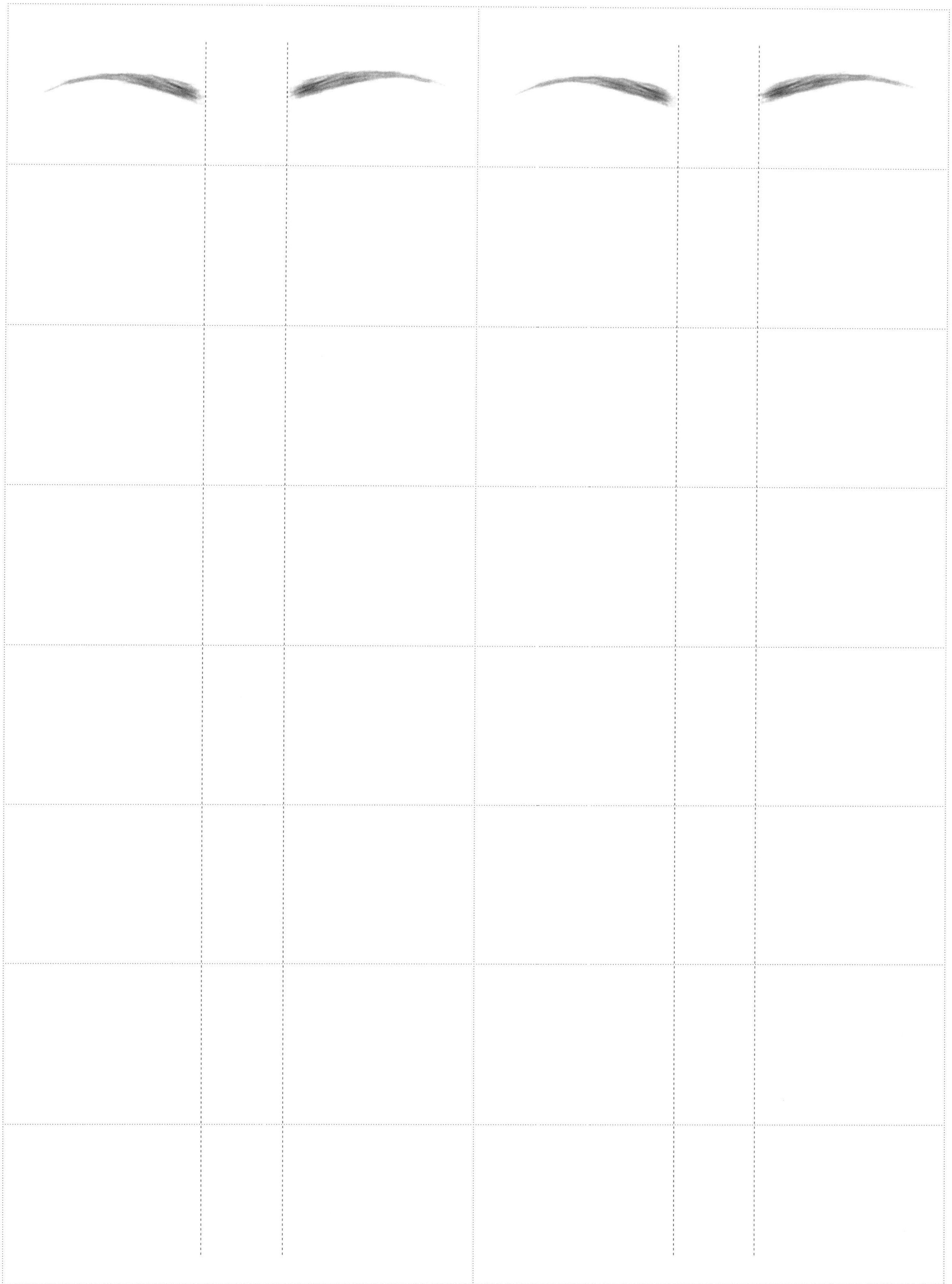

여자 얼굴 정면(눈썹연습)

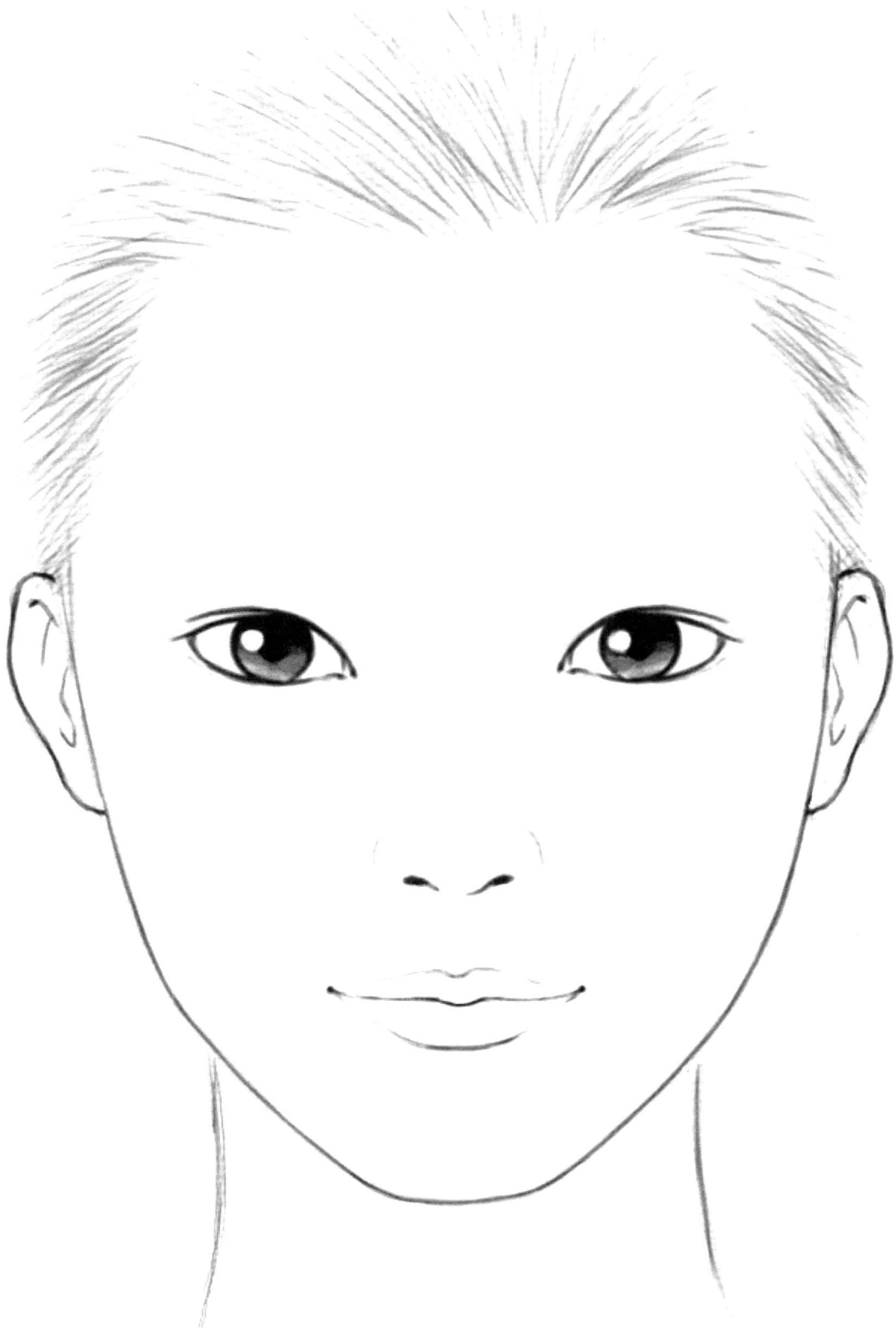

Make-up Title :		Skin Color :	
Eye-Shadow			
High-Light Color	Main Color	Point Color	Accent Color
Cheek Color		Lip Color	

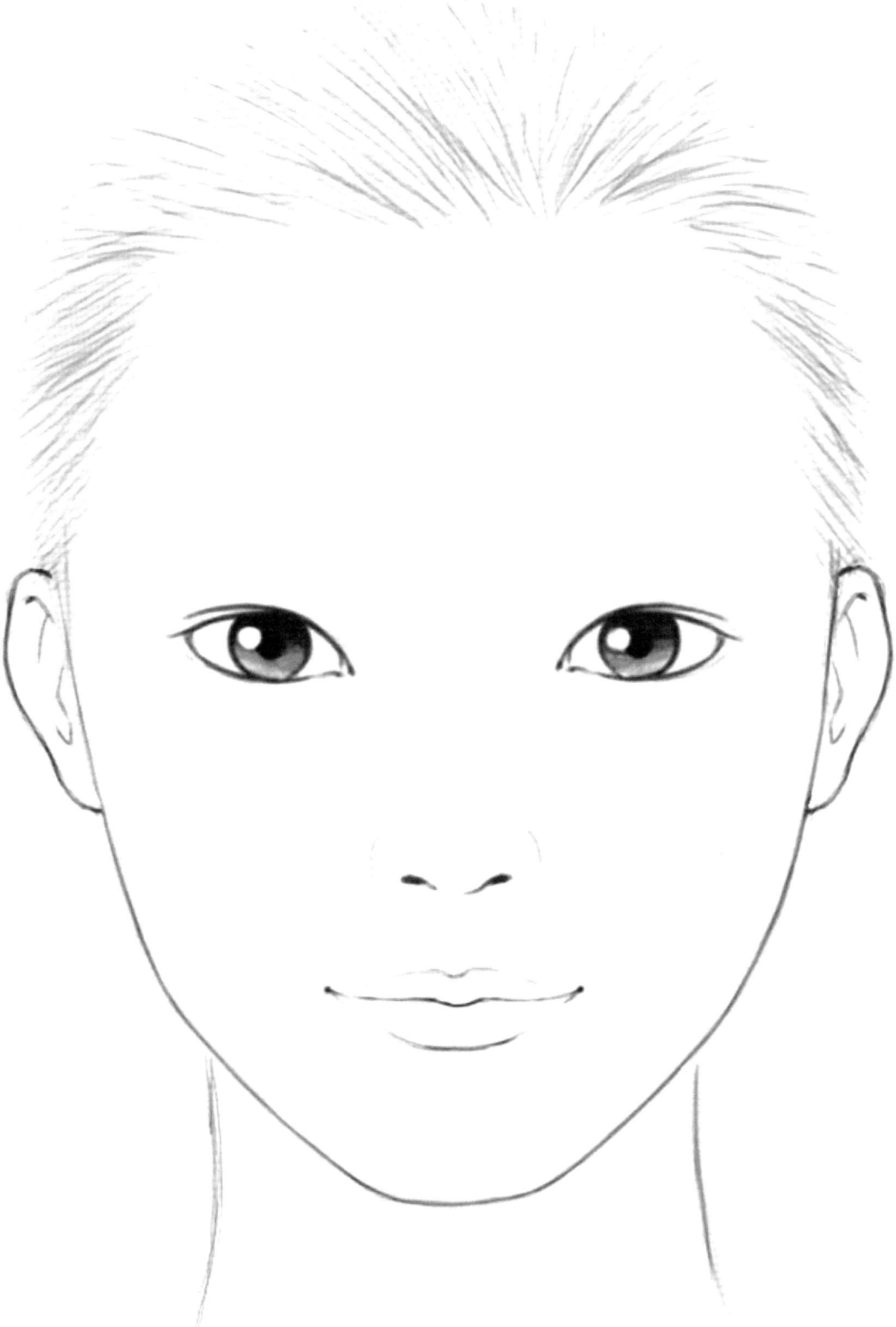

Make-up Title :		Skin Color :	
Eye-Shadow			
High-Light Color	Main Color	Point Color	Accent Color
Cheek Color		Lip Color	

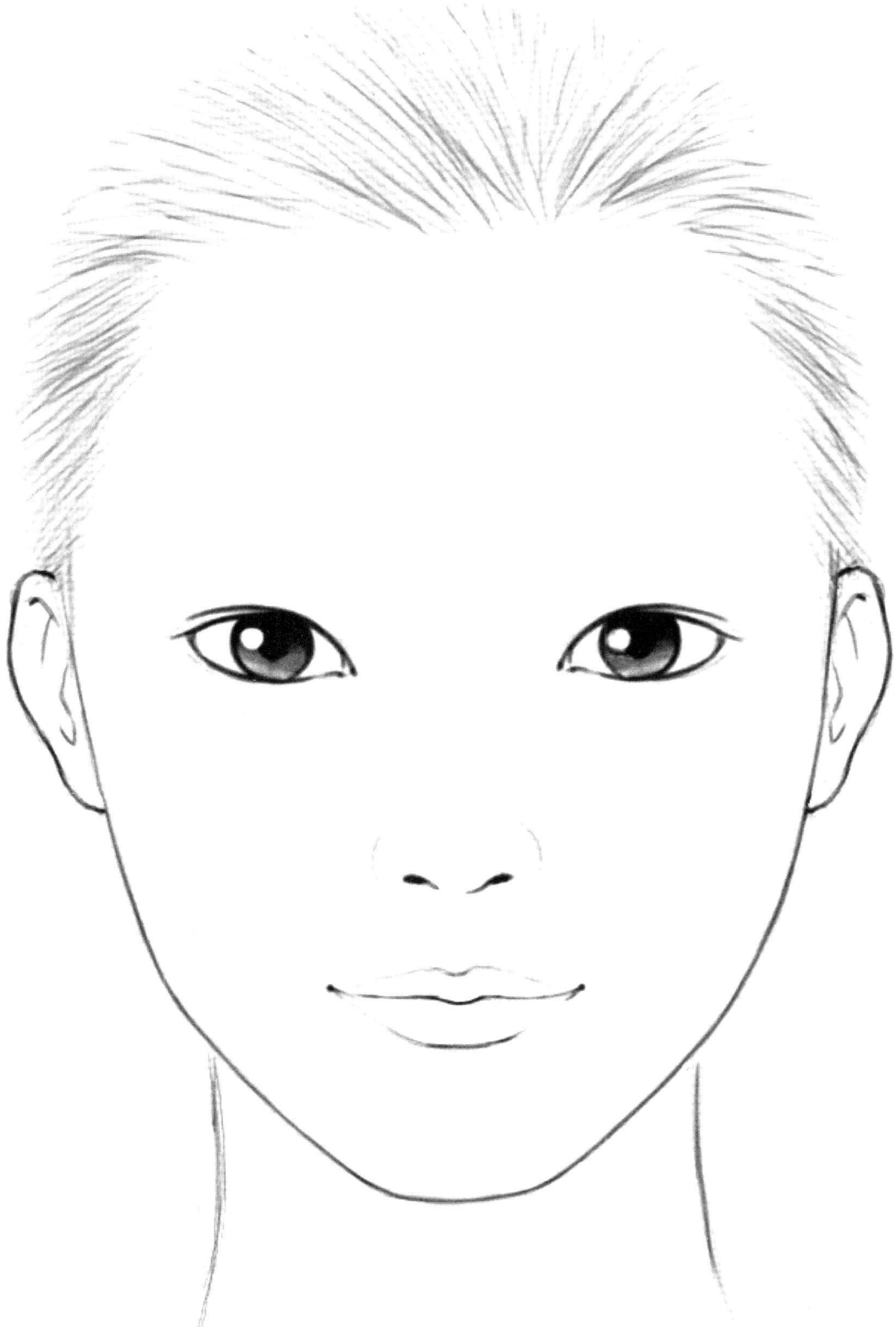

Make-up Title :		Skin Color :	
Eye-Shadow			
High-Light Color	Main Color	Point Color	Accent Color
Cheek Color		Lip Color	

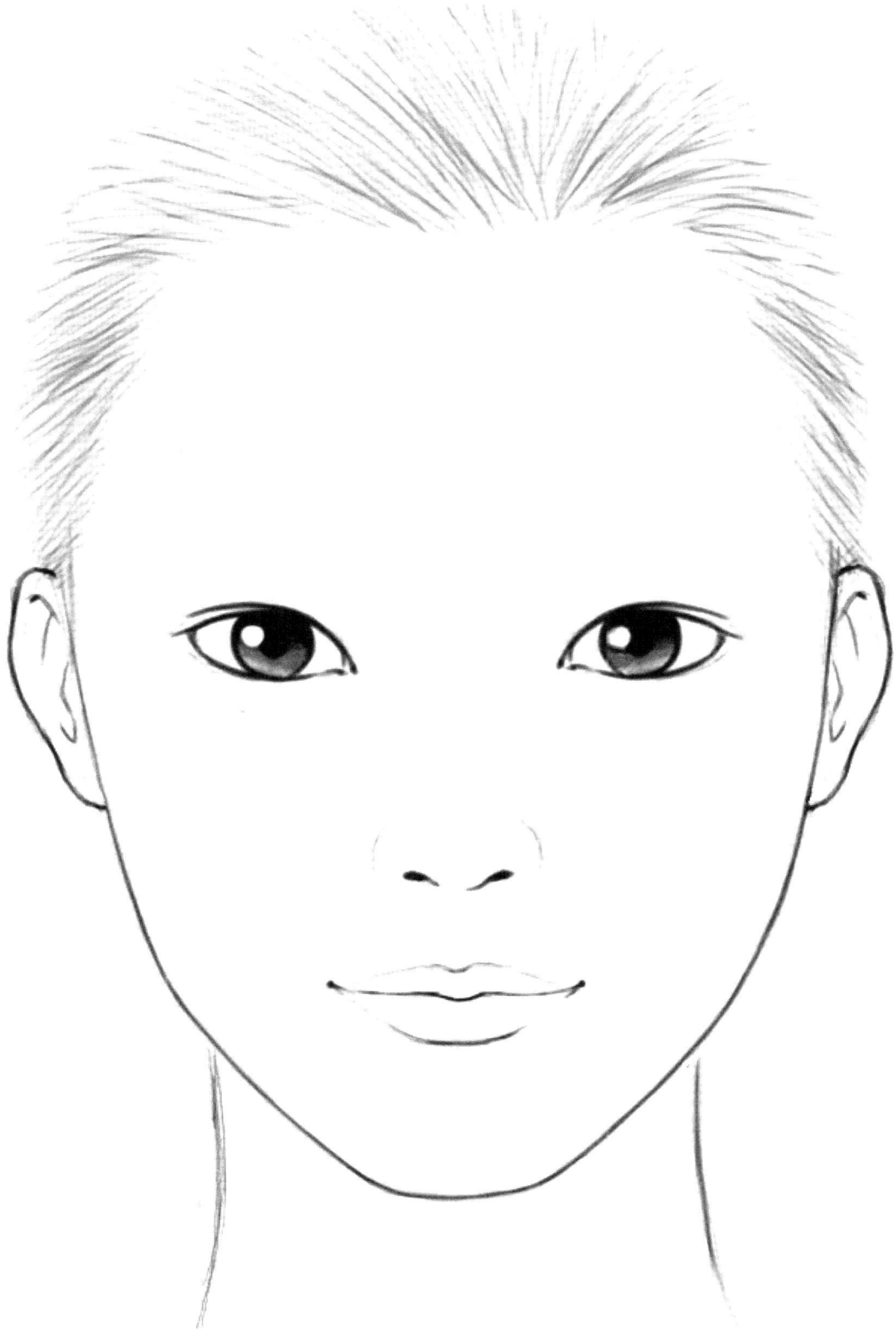

Make-up Title :		Skin Color :	
Eye-Shadow			
High-Light Color	Main Color	Point Color	Accent Color
Cheek Color		Lip Color	

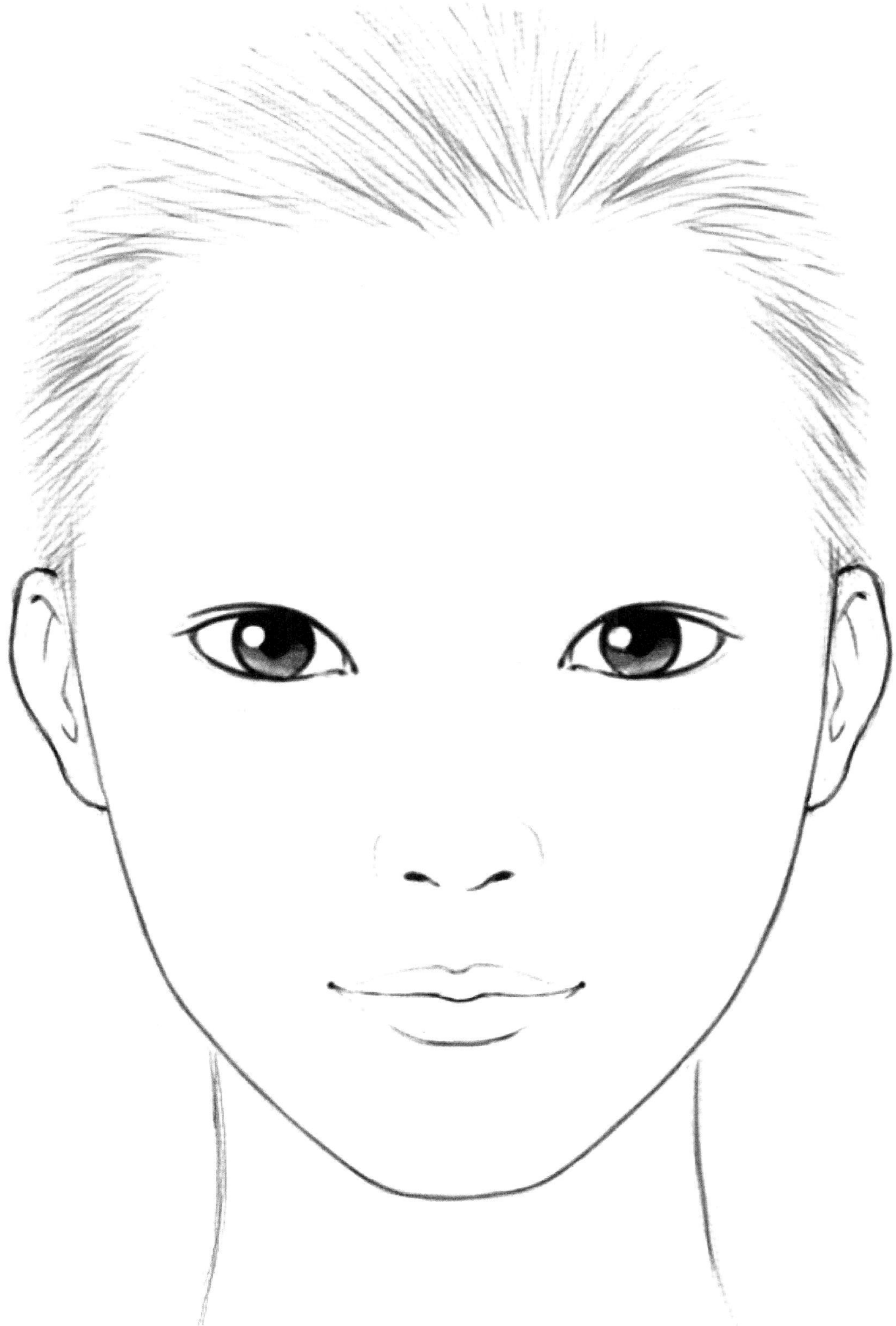

Make-up Title :		Skin Color :	
Eye-Shadow			
High-Light Color	Main Color	Point Color	Accent Color
Cheek Color		Lip Color	

Chapter 10

Eye-Shadow Style

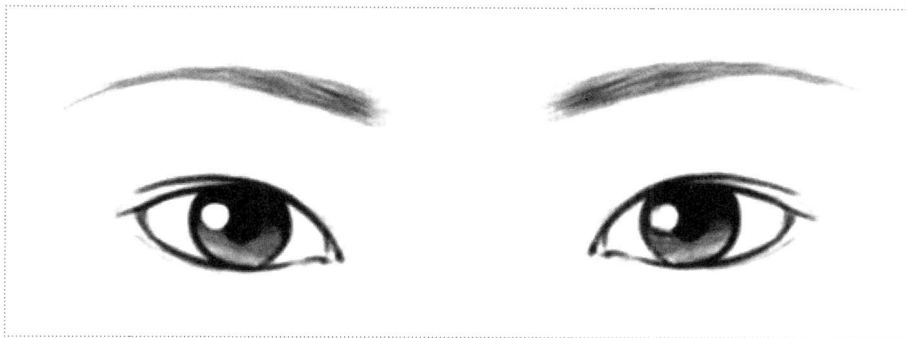

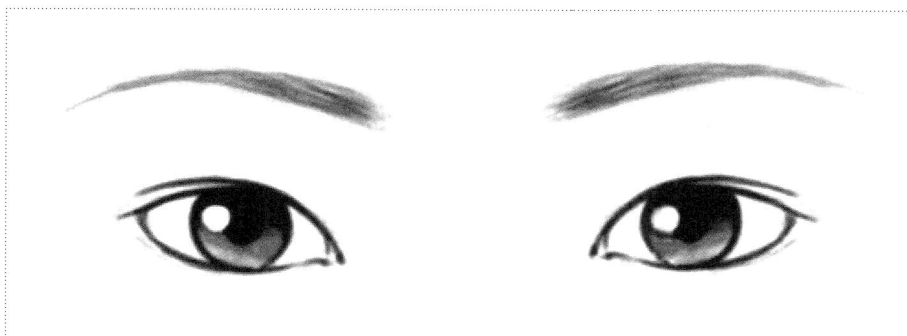

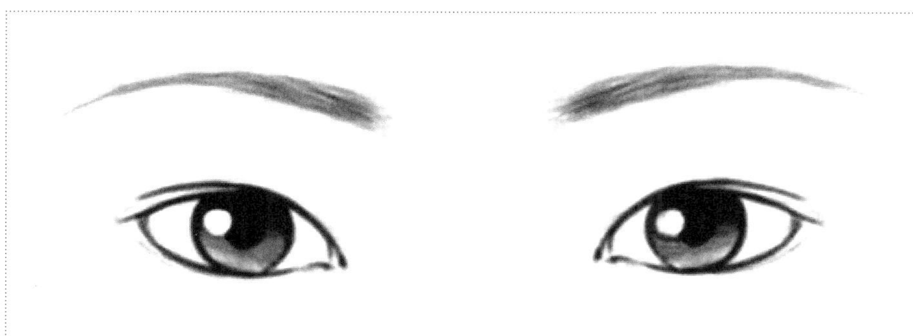

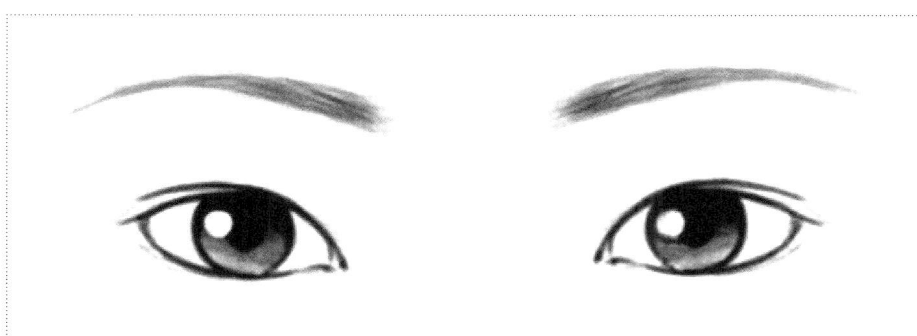

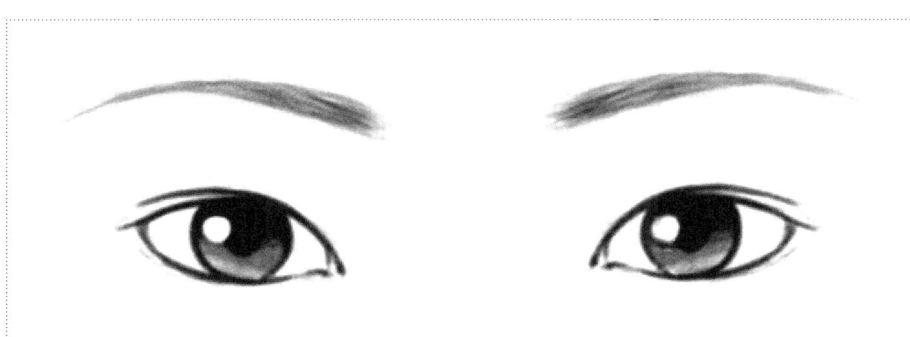

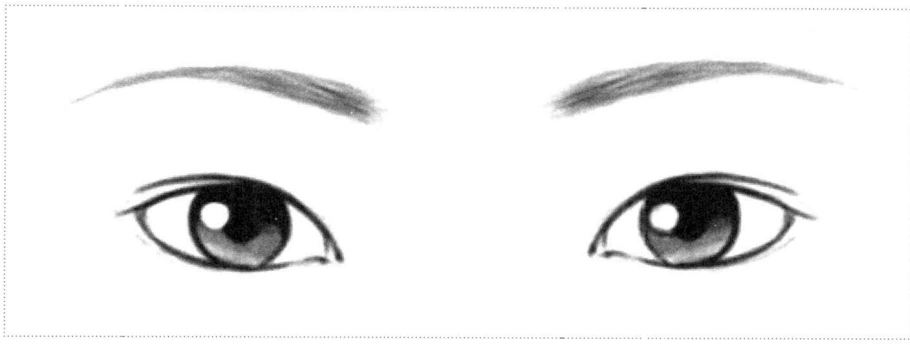

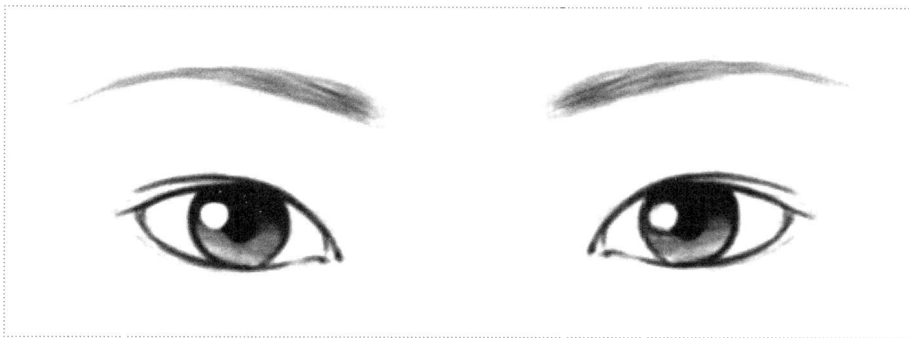

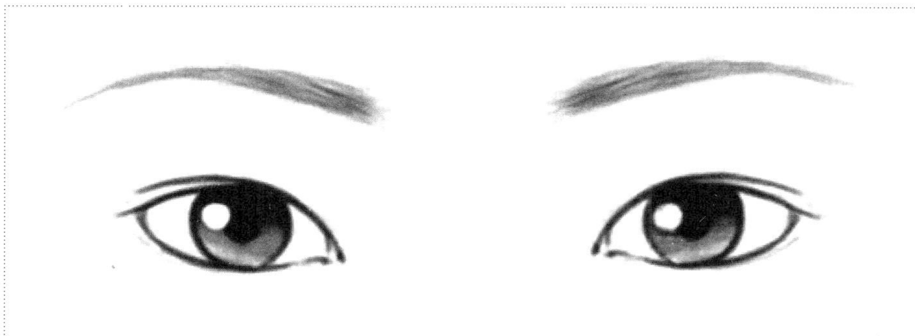

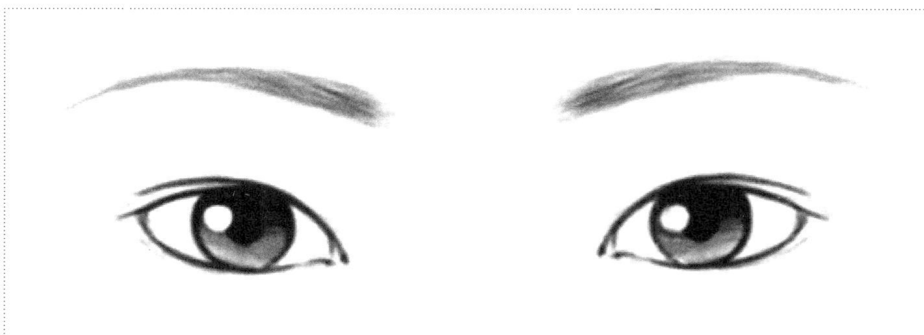

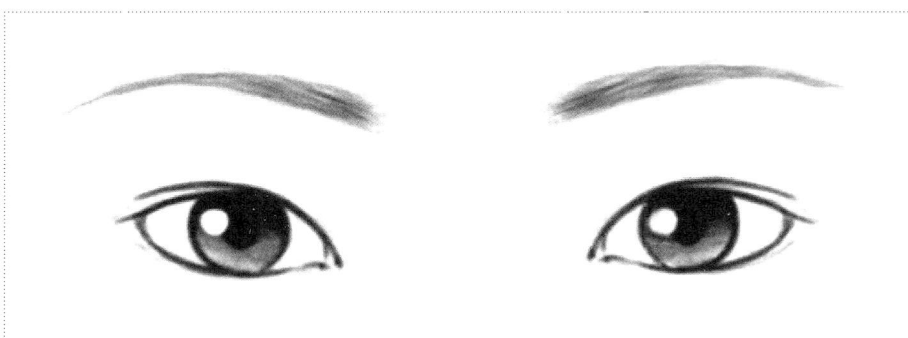

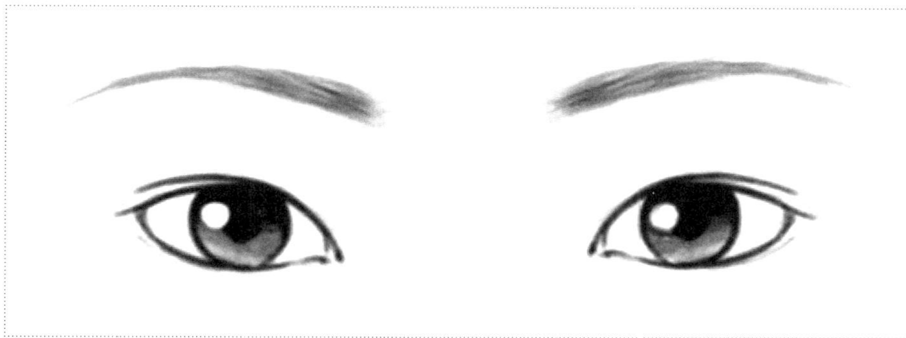
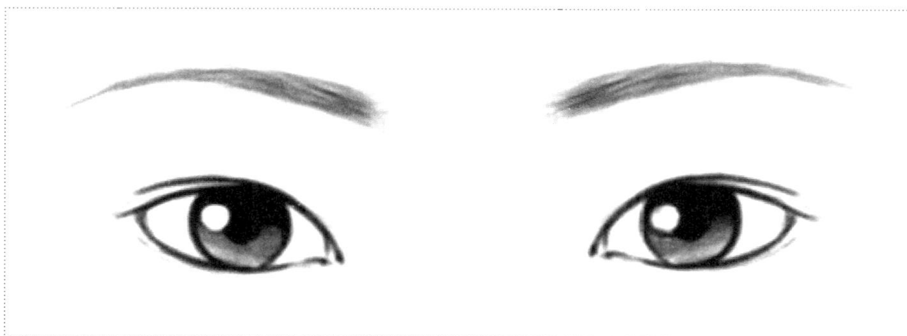
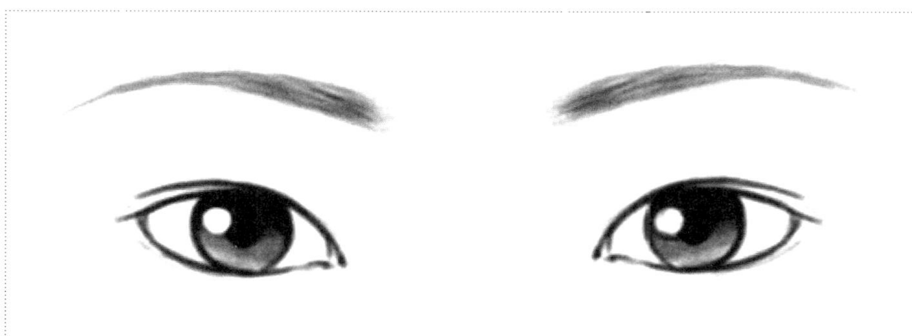
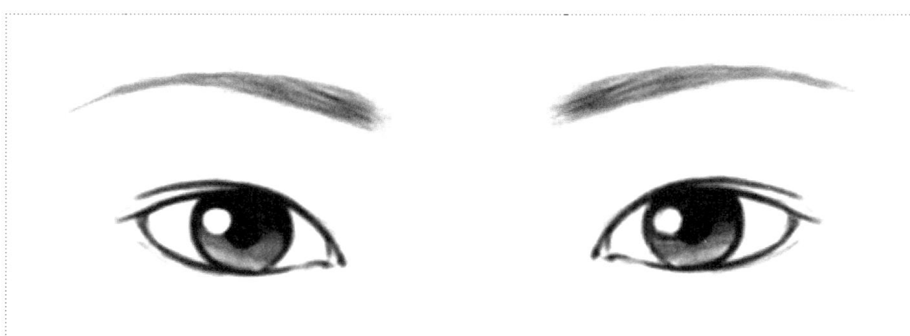
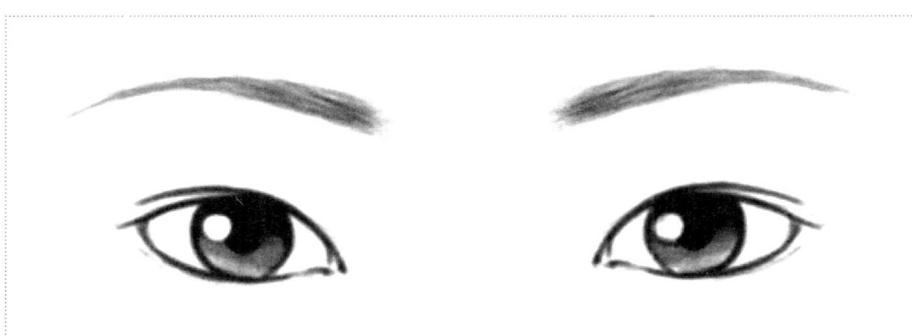

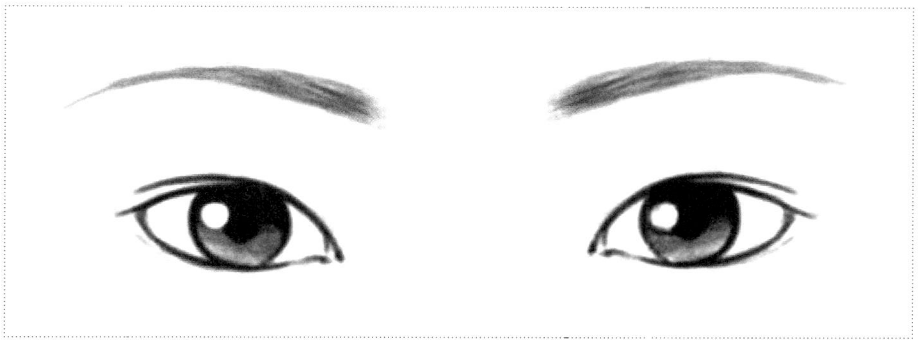
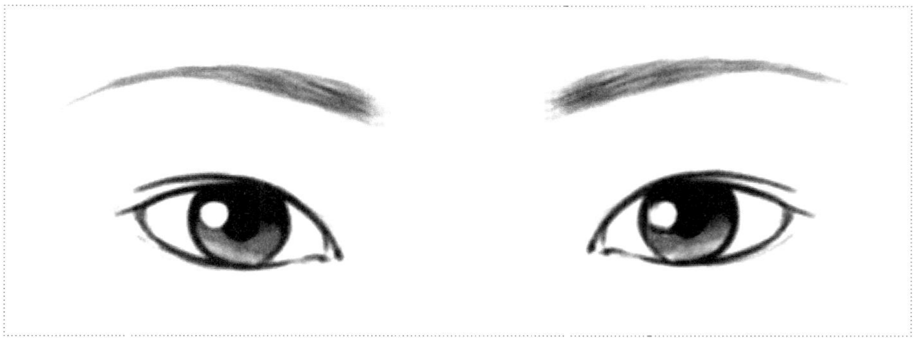
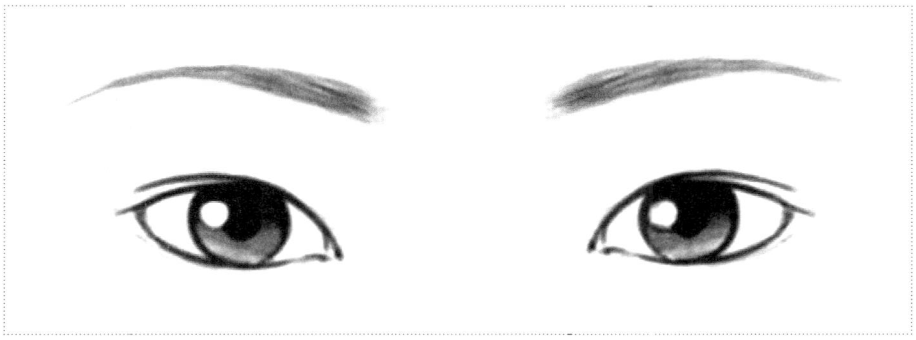
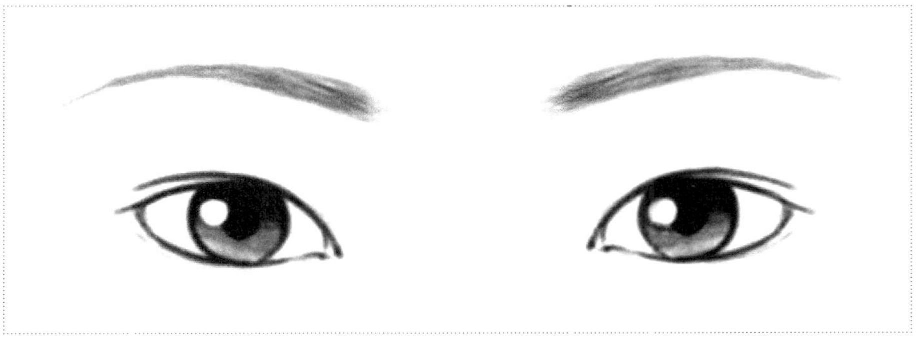
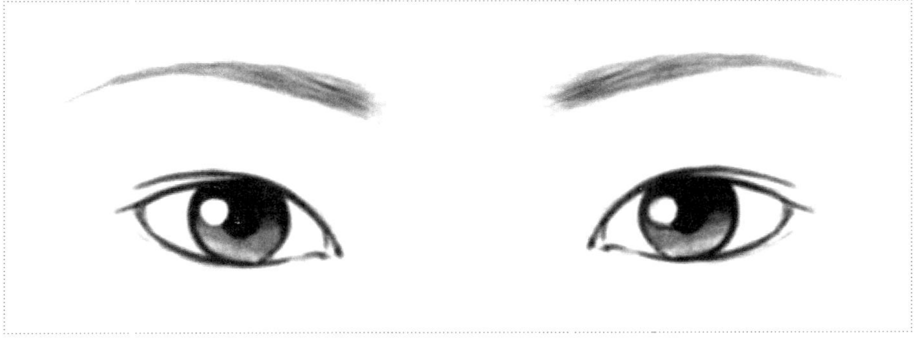

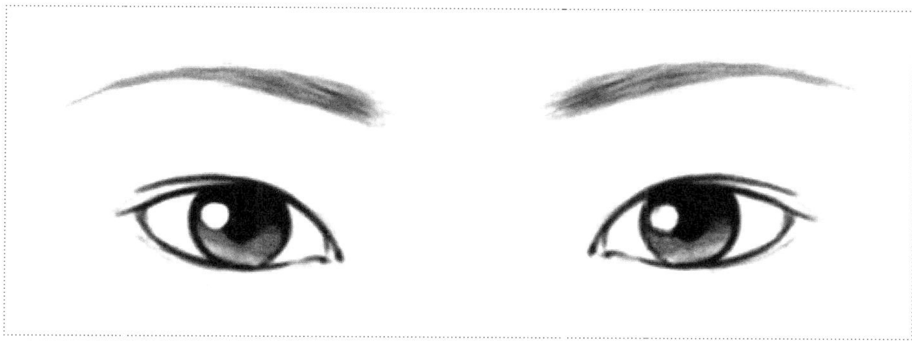

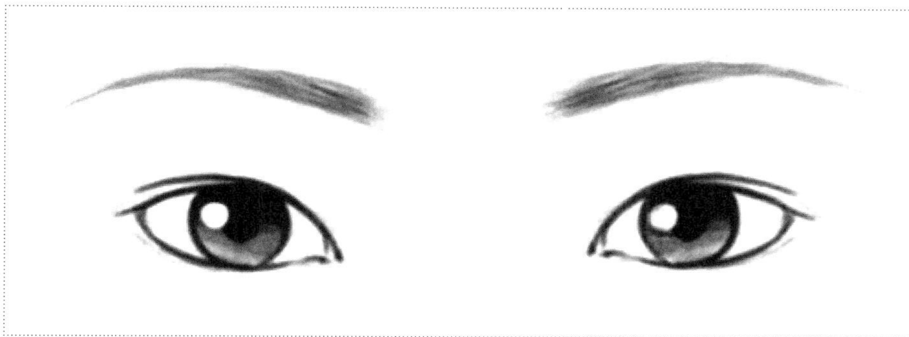

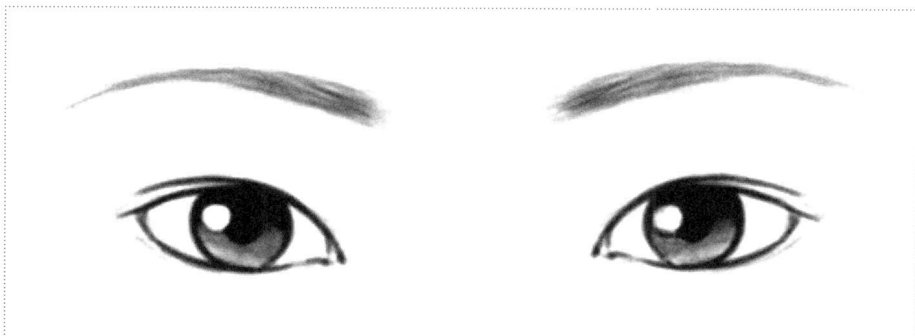

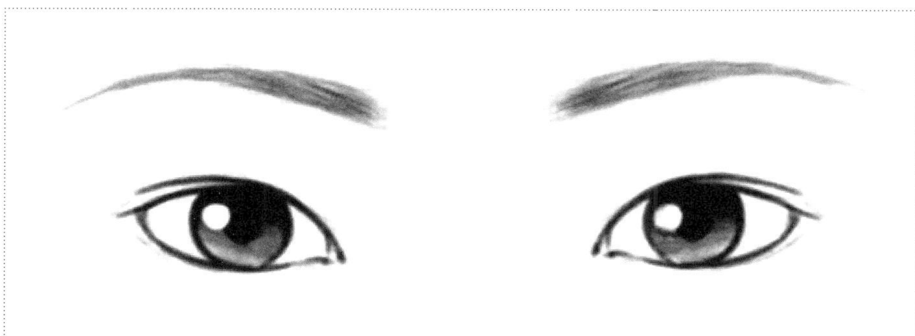

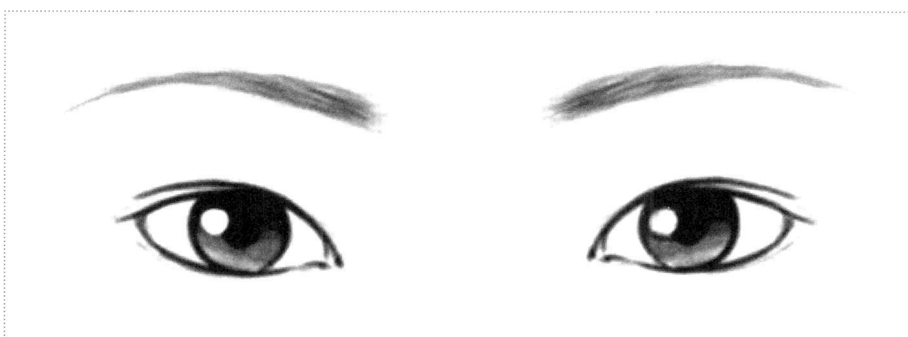

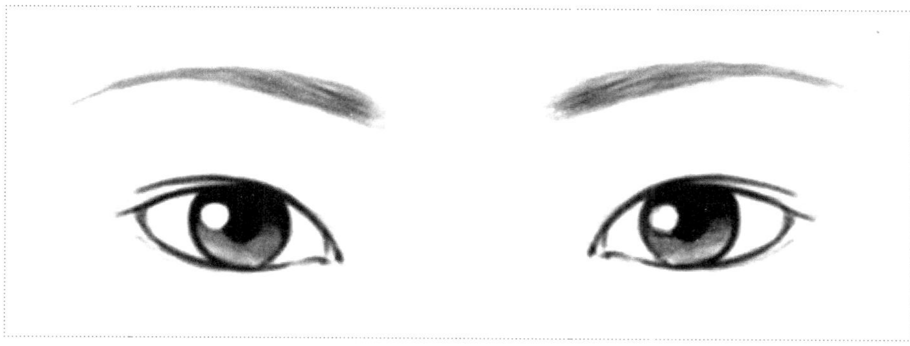

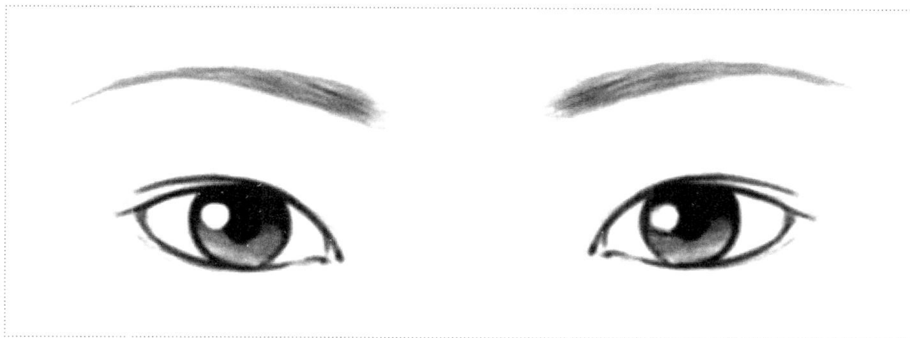

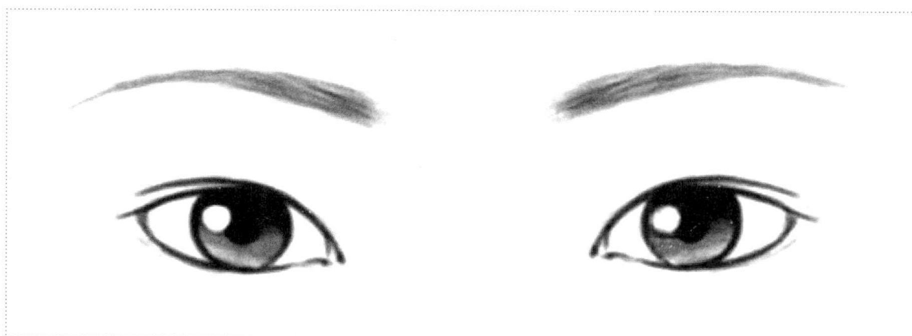

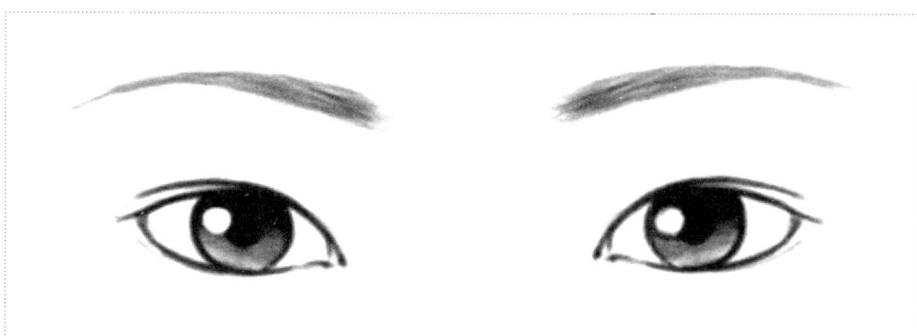

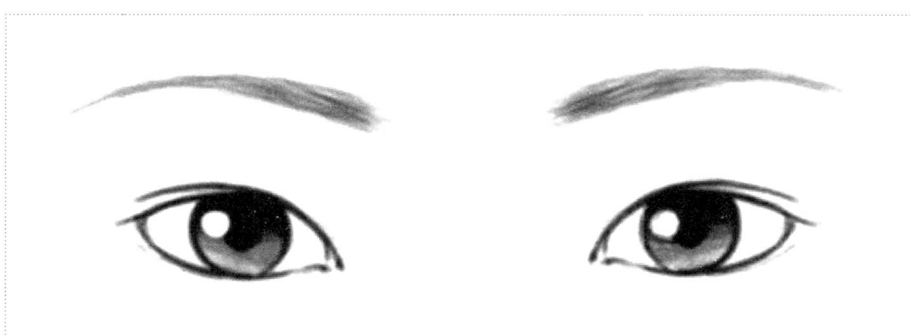

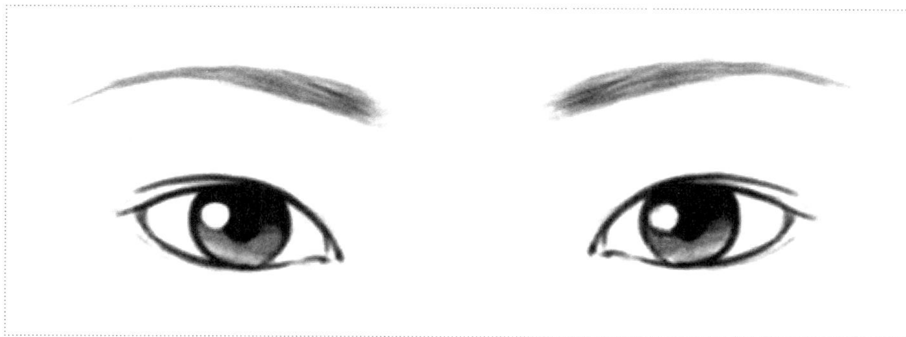

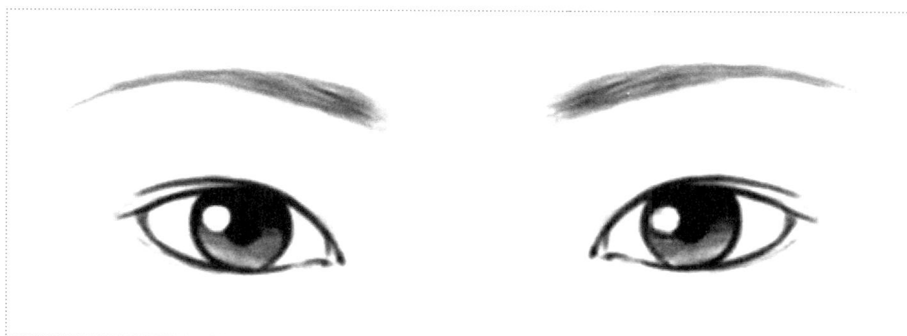

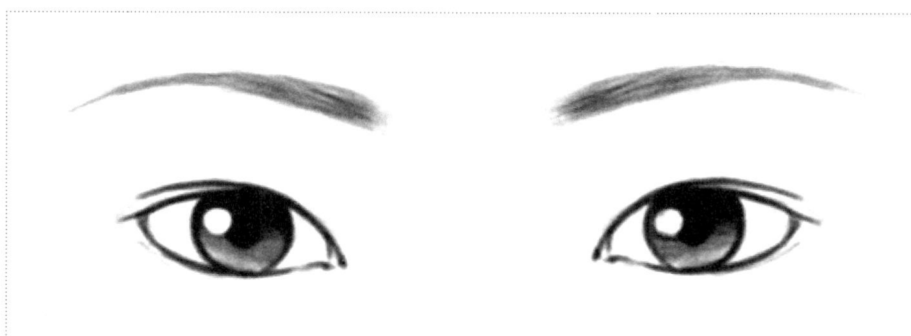

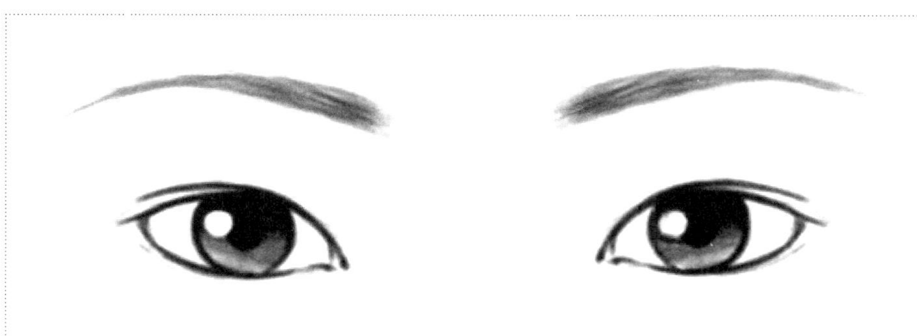

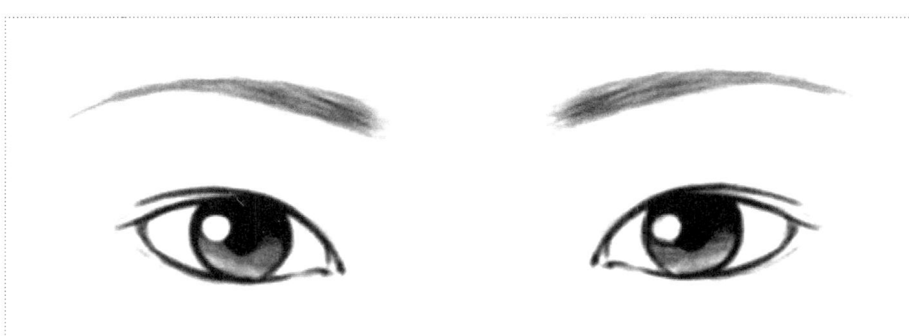

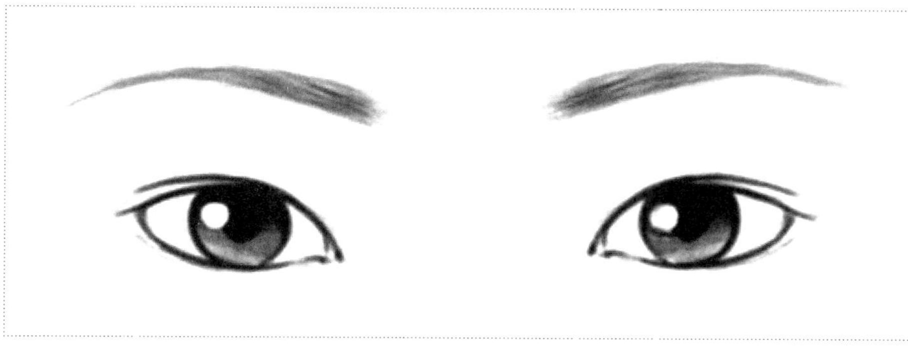

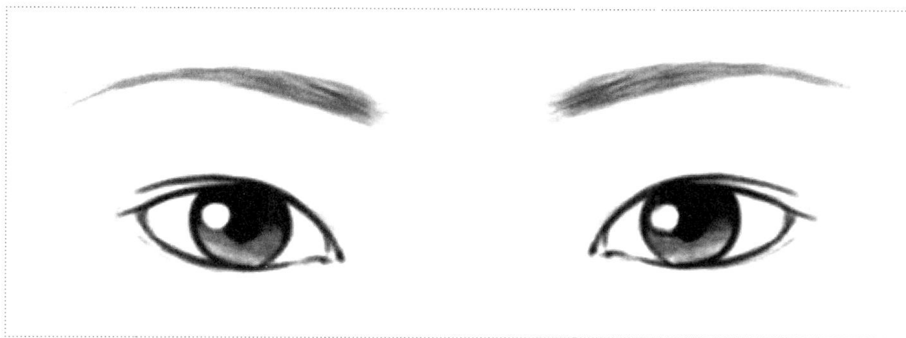

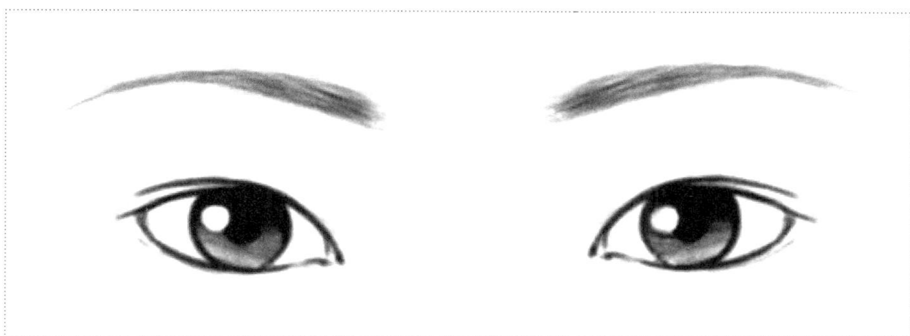

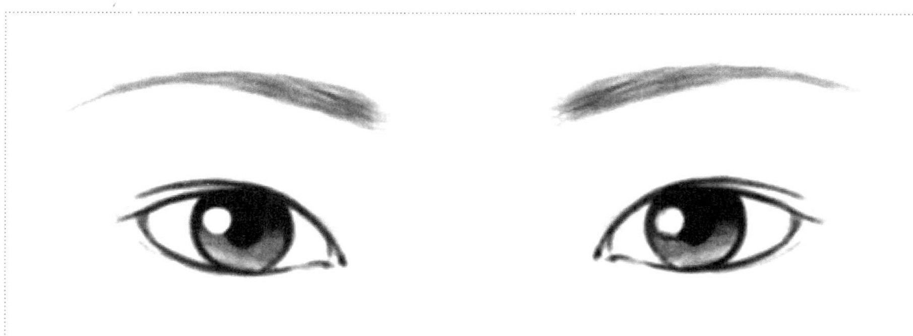

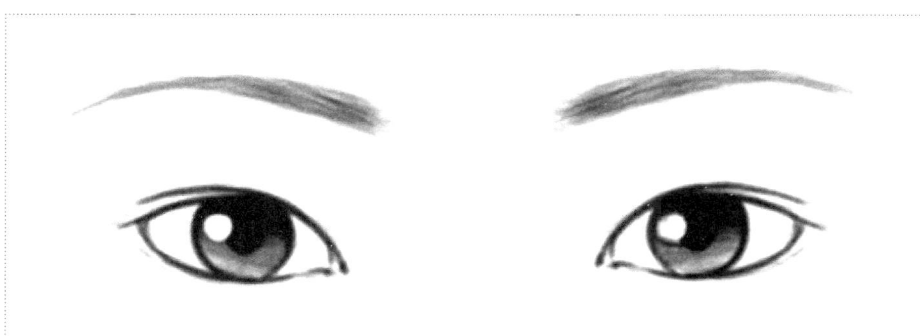

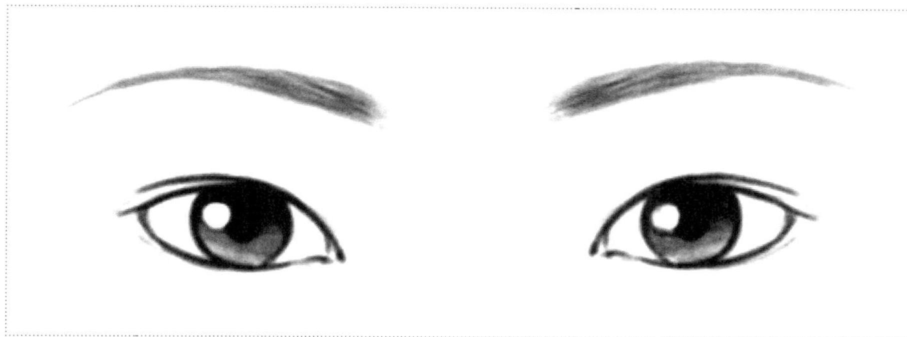
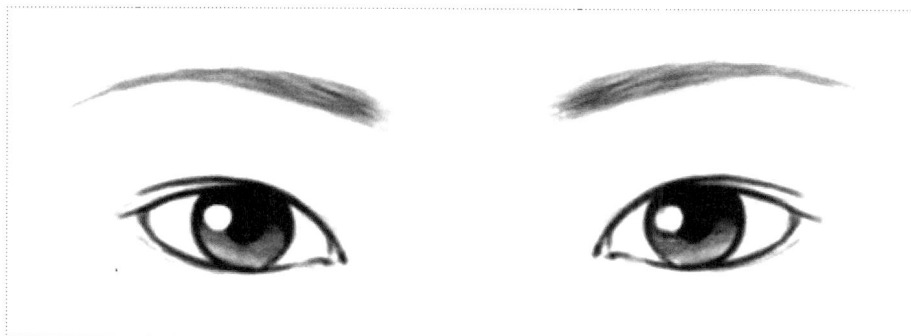
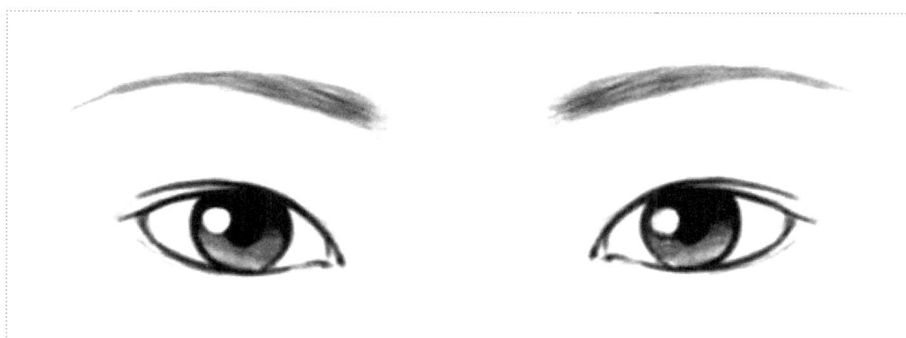
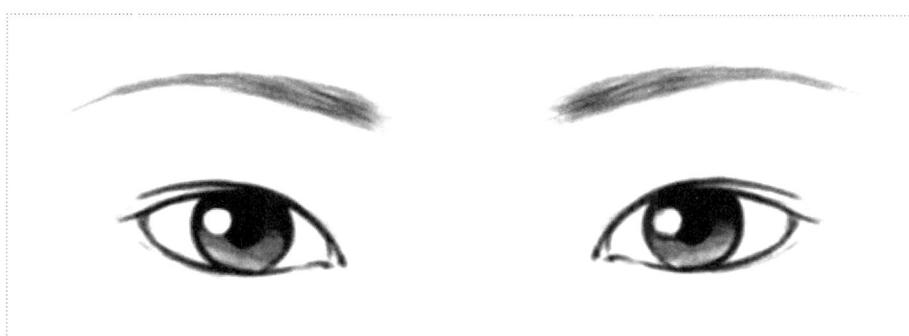
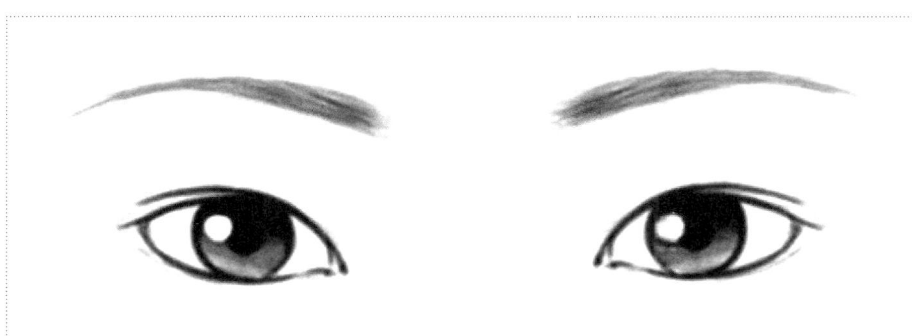

Chapter 11

여자 얼굴 정면

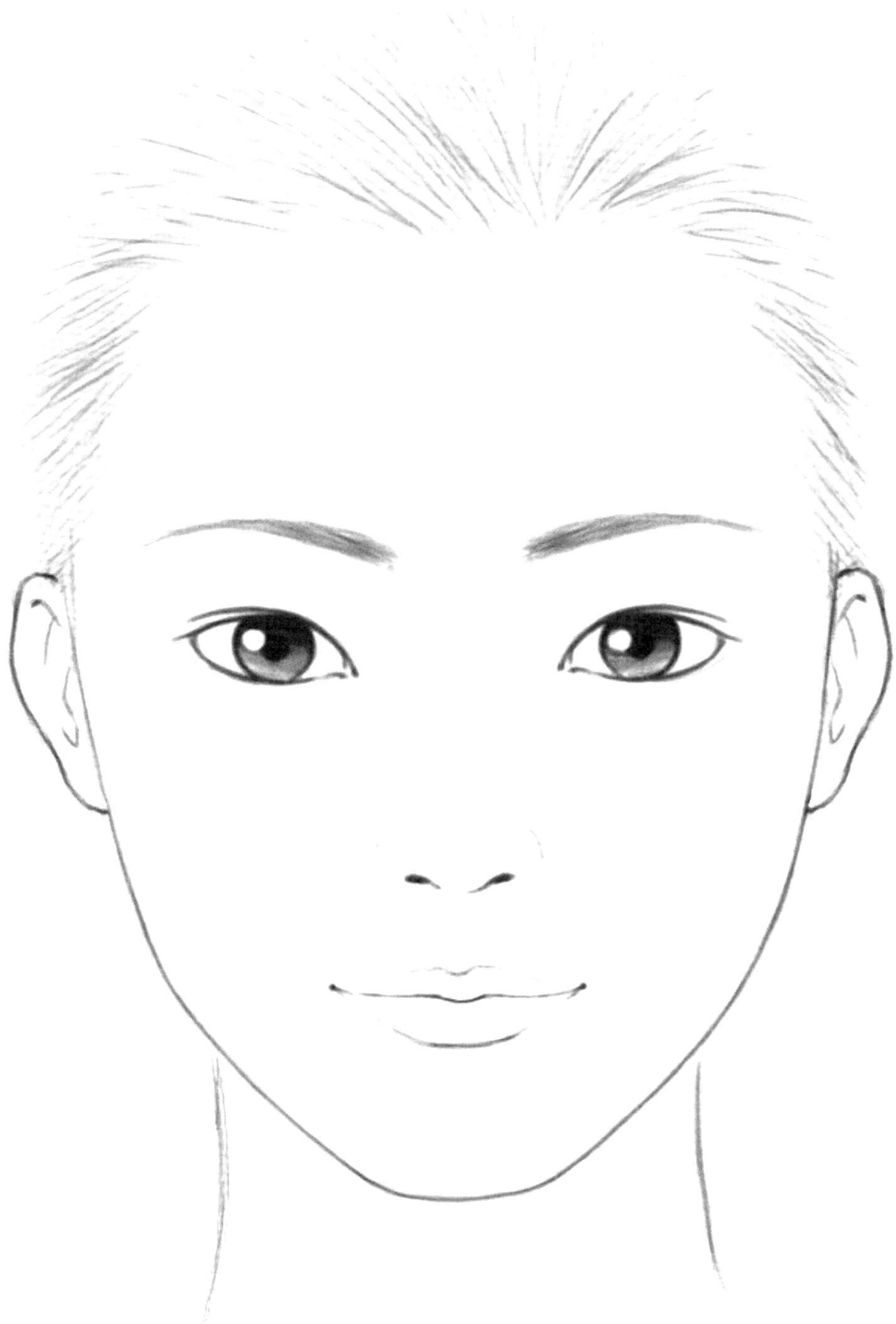

Make-up Title :		Skin Color :	
Eye-Shadow			
High-Light Color	Main Color	Point Color	Accent Color
Cheek Color		Lip Color	

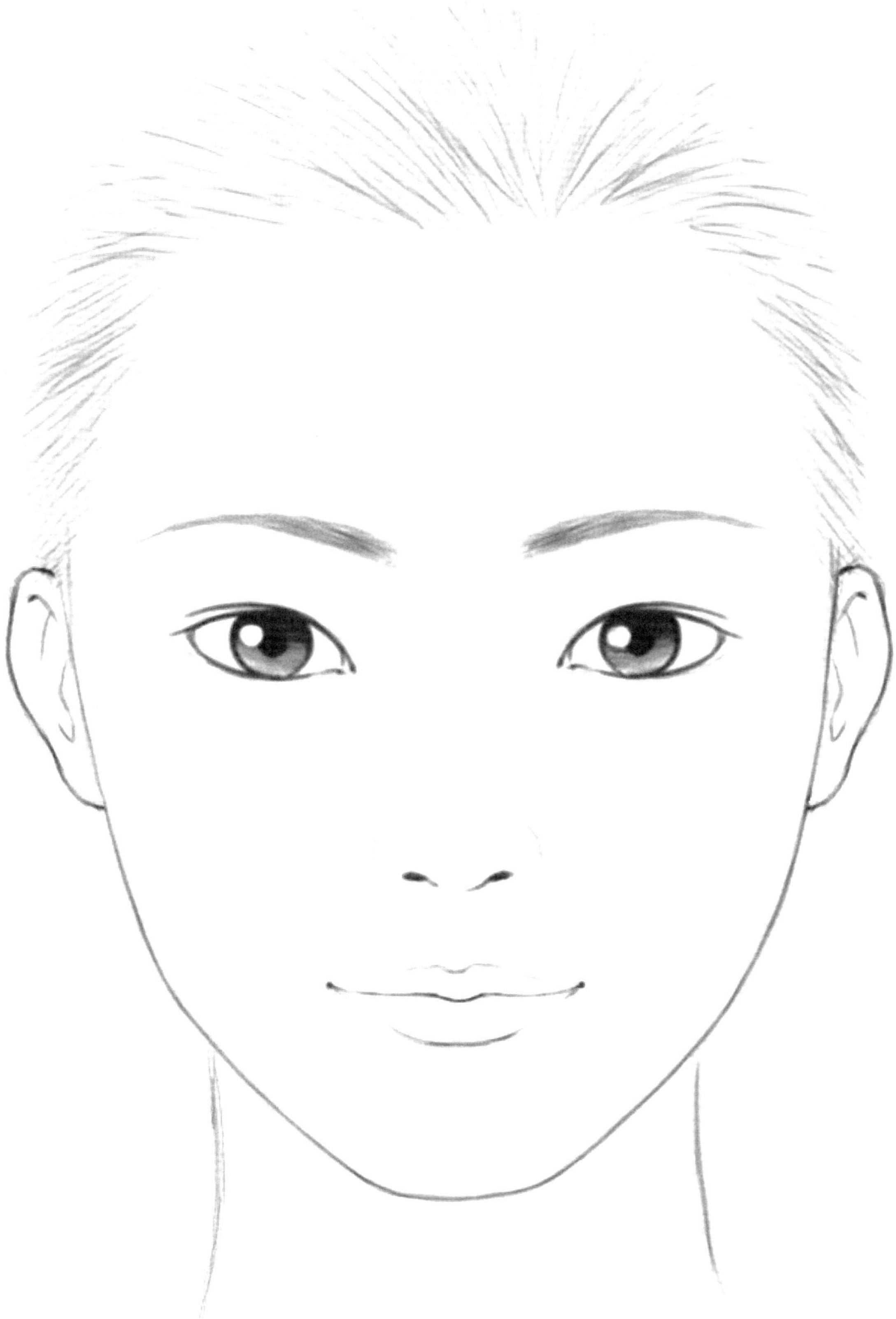

Make-up Title :		Skin Color :	
Eye-Shadow			
High-Light Color	Main Color	Point Color	Accent Color
Cheek Color		Lip Color	

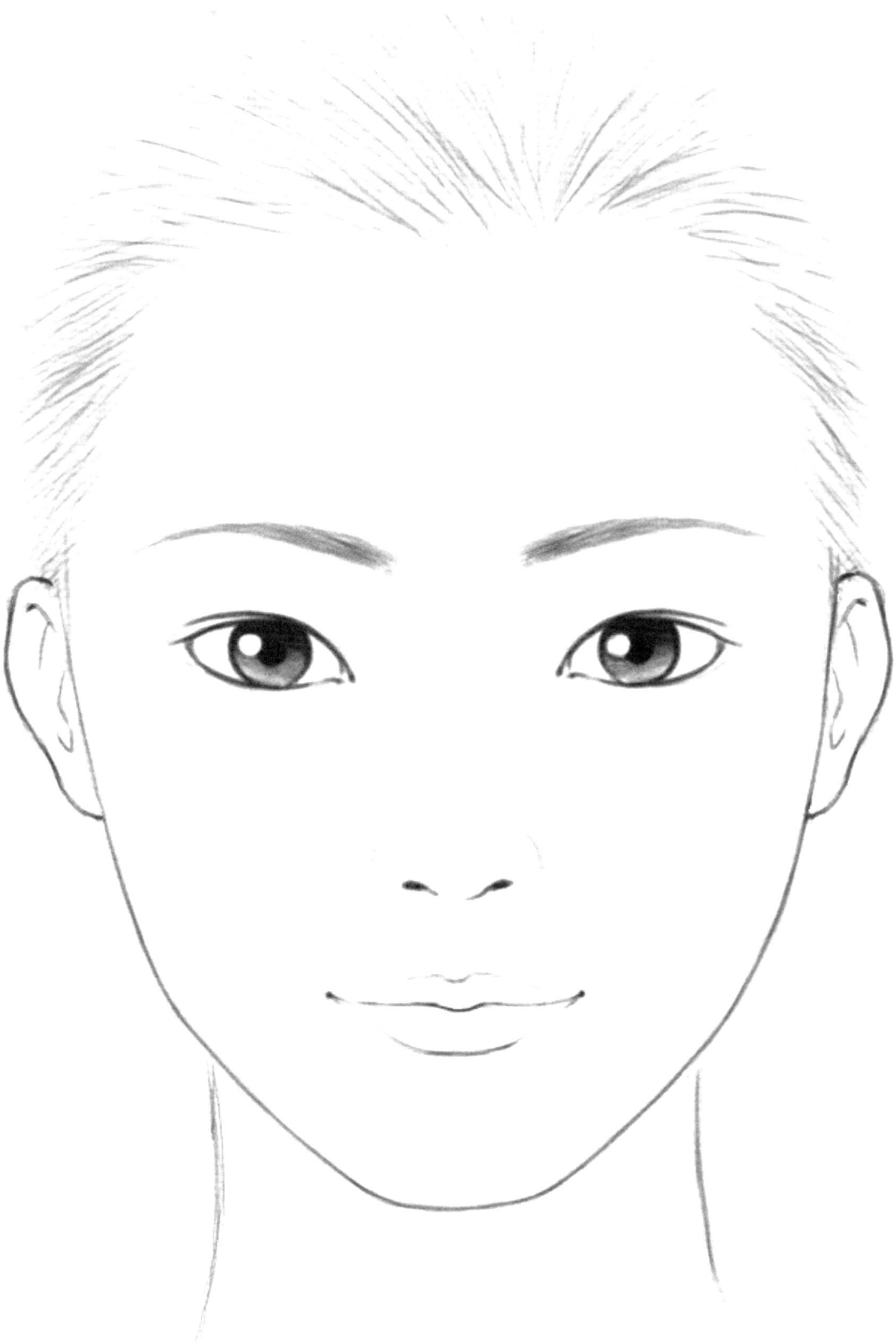

Make-up Title :		Skin Color :	
Eye-Shadow			
High-Light Color	Main Color	Point Color	Accent Color
Cheek Color		Lip Color	

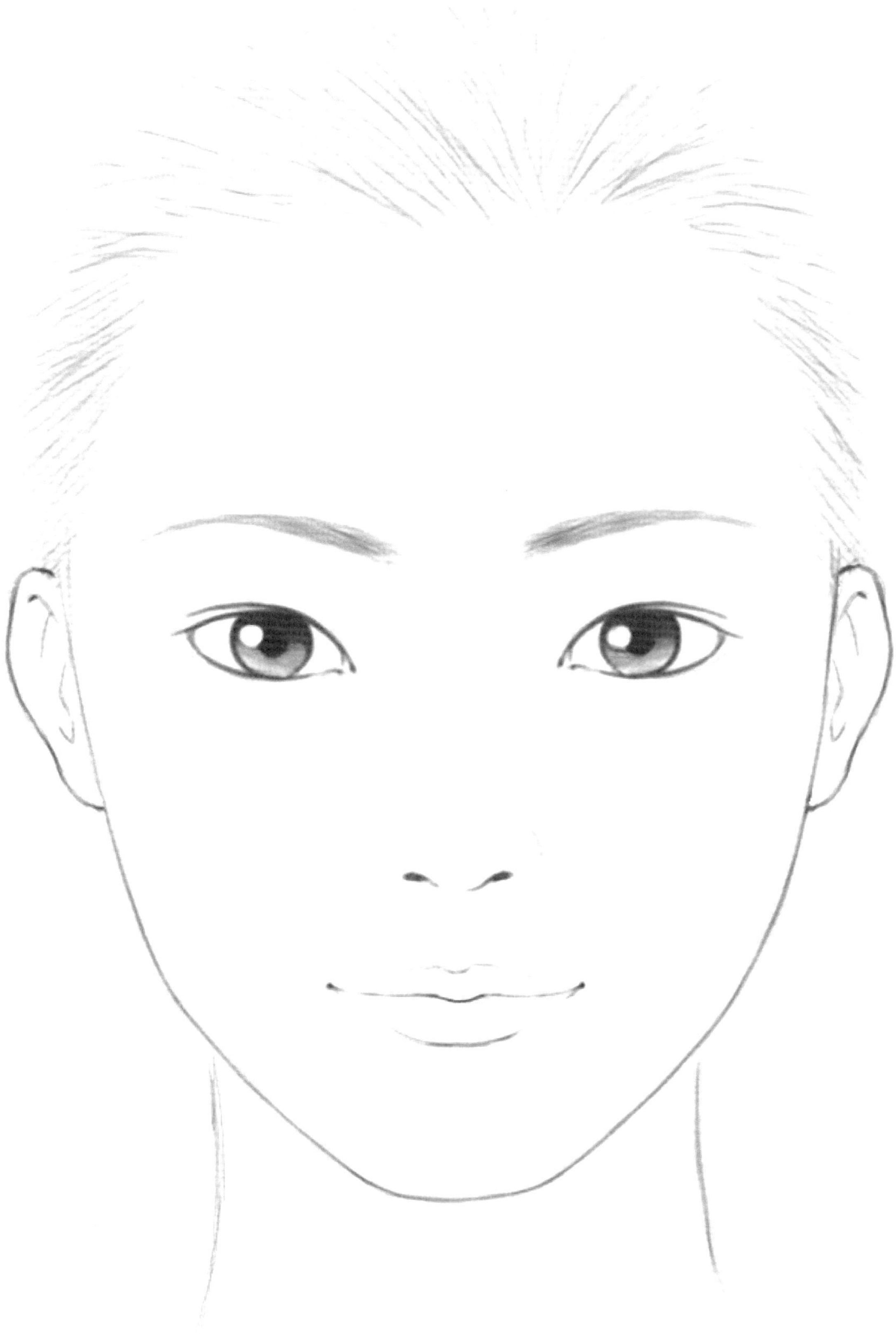

Make-up Title :		Skin Color :	
Eye-Shadow			
High-Light Color	Main Color	Point Color	Accent Color
Cheek Color		Lip Color	

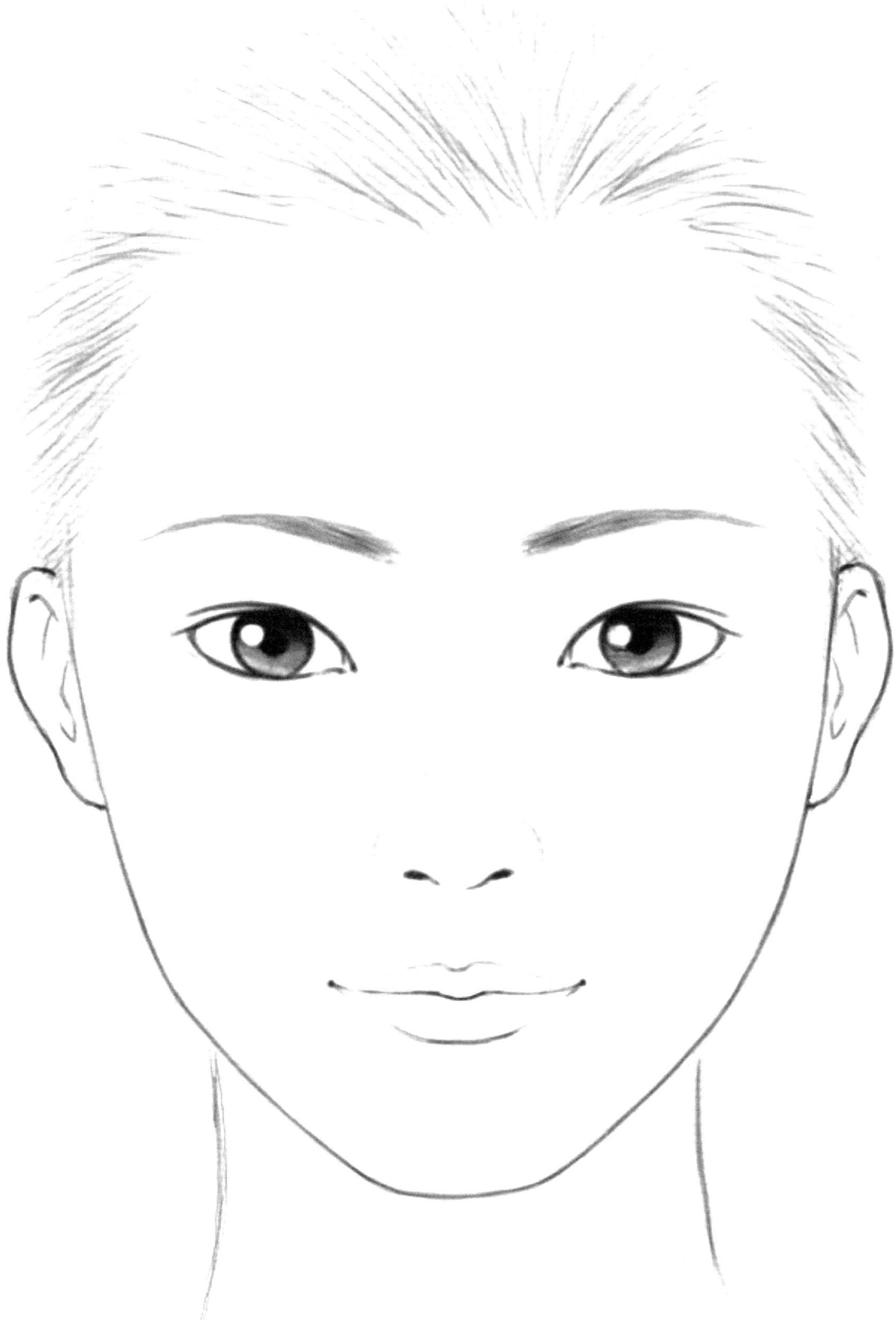

Make-up Title :		Skin Color :	
Eye-Shadow			
High-Light Color	Main Color	Point Color	Accent Color
Cheek Color		Lip Color	

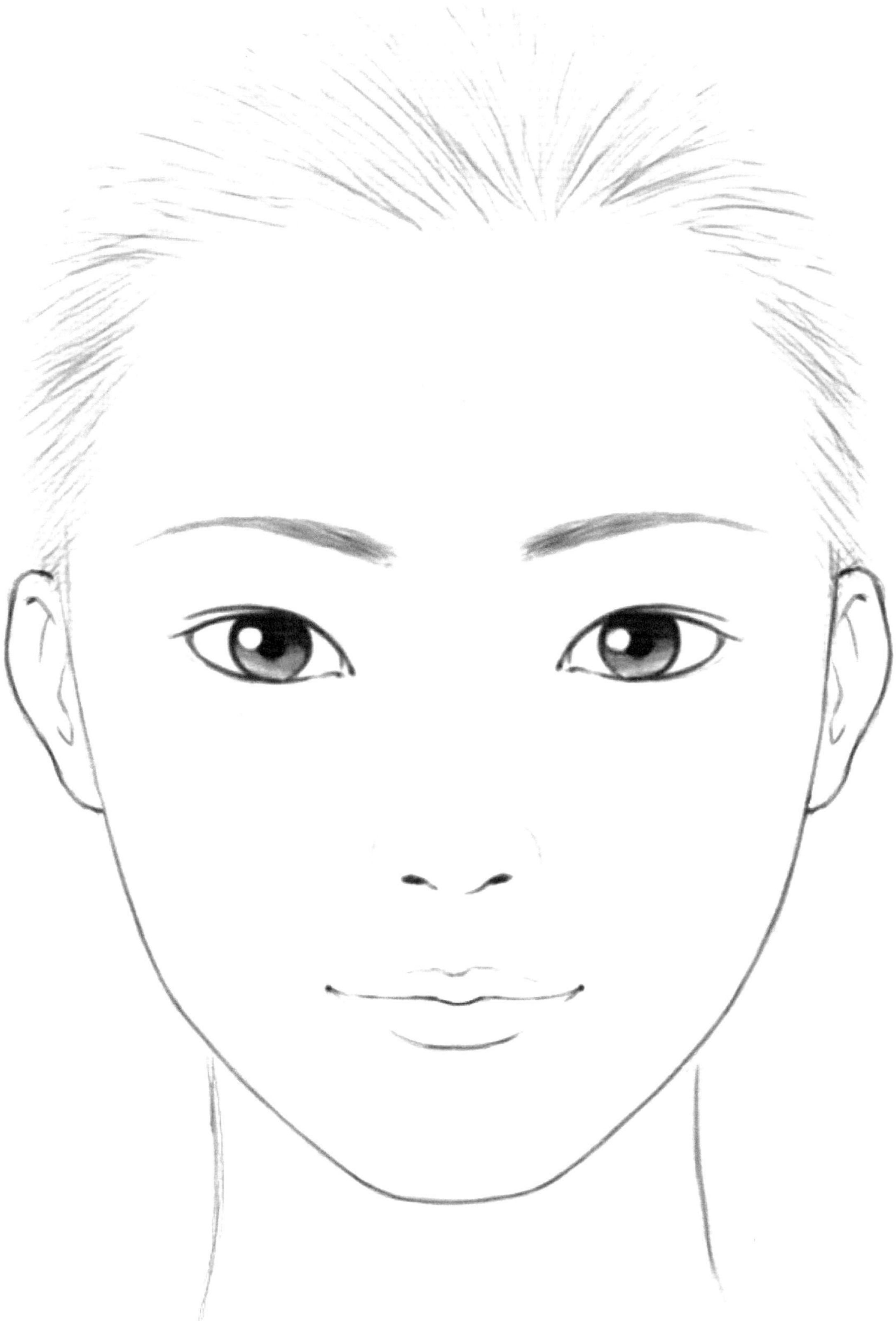

Make-up Title :		Skin Color :	
Eye-Shadow			
High-Light Color	Main Color	Point Color	Accent Color
Cheek Color		Lip Color	

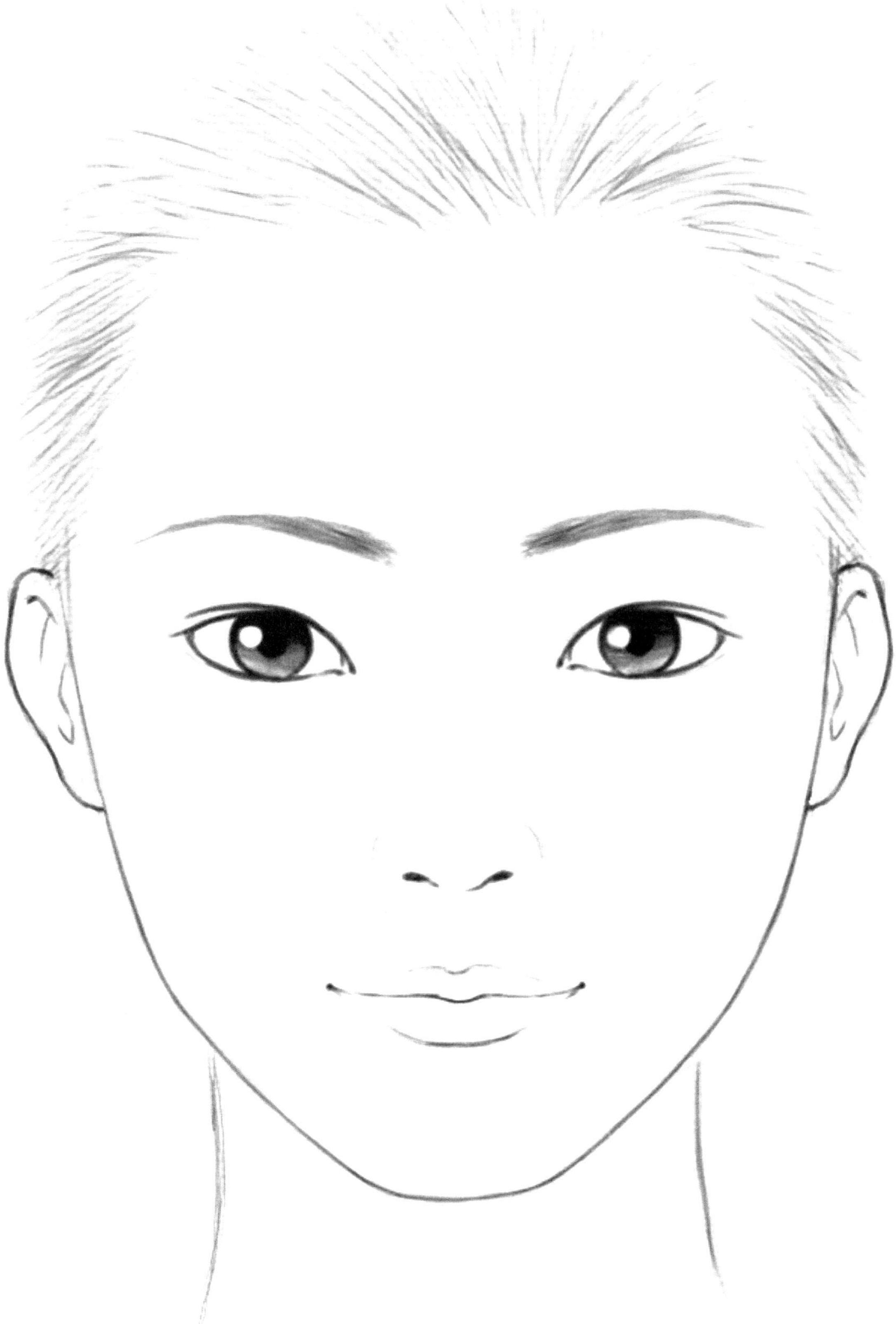

Make-up Title :		Skin Color :	
Eye-Shadow			
High-Light Color	Main Color	Point Color	Accent Color
Cheek Color		Lip Color	

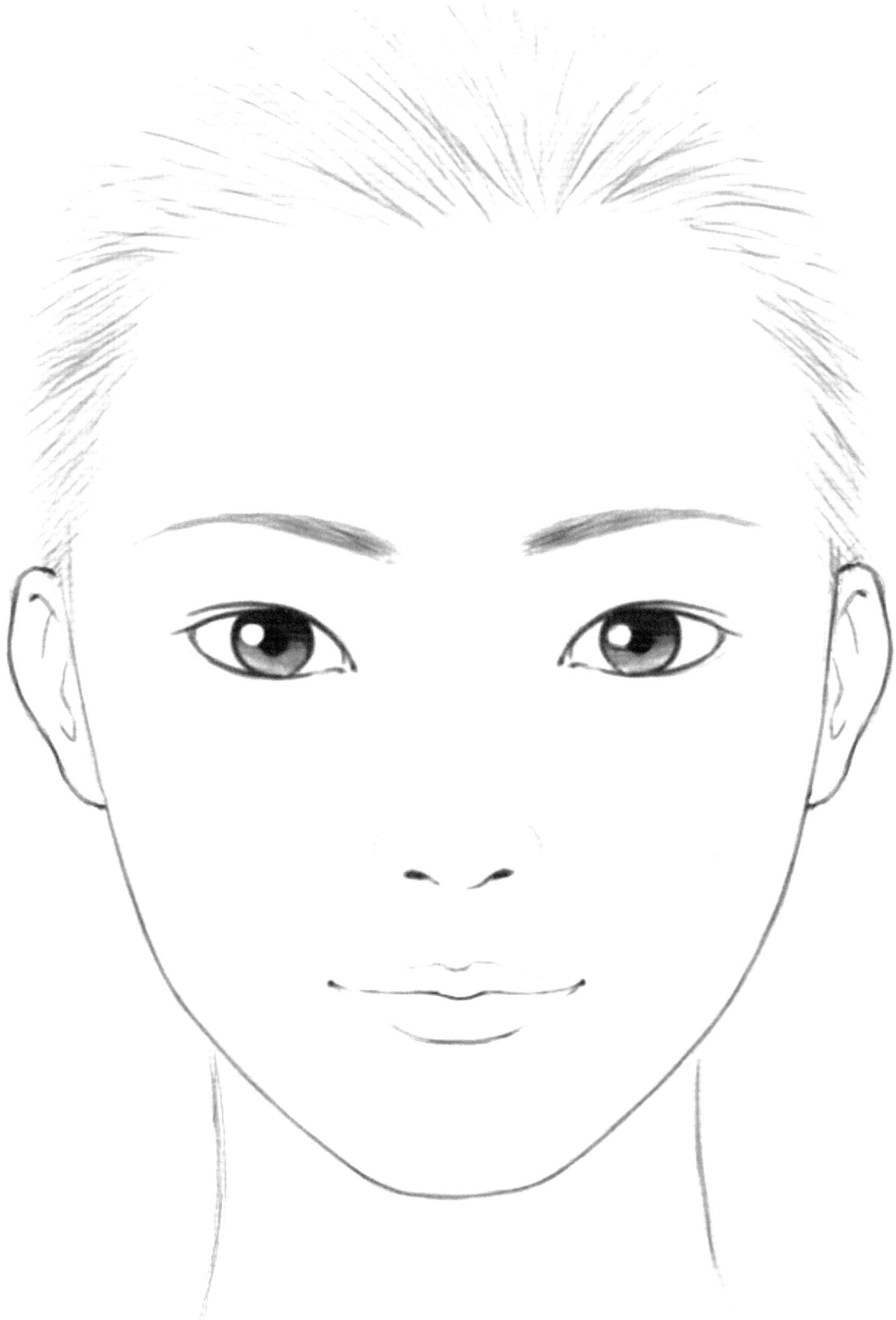

Make-up Title :		Skin Color :	
Eye-Shadow			
High-Light Color	Main Color	Point Color	Accent Color
Cheek Color		Lip Color	

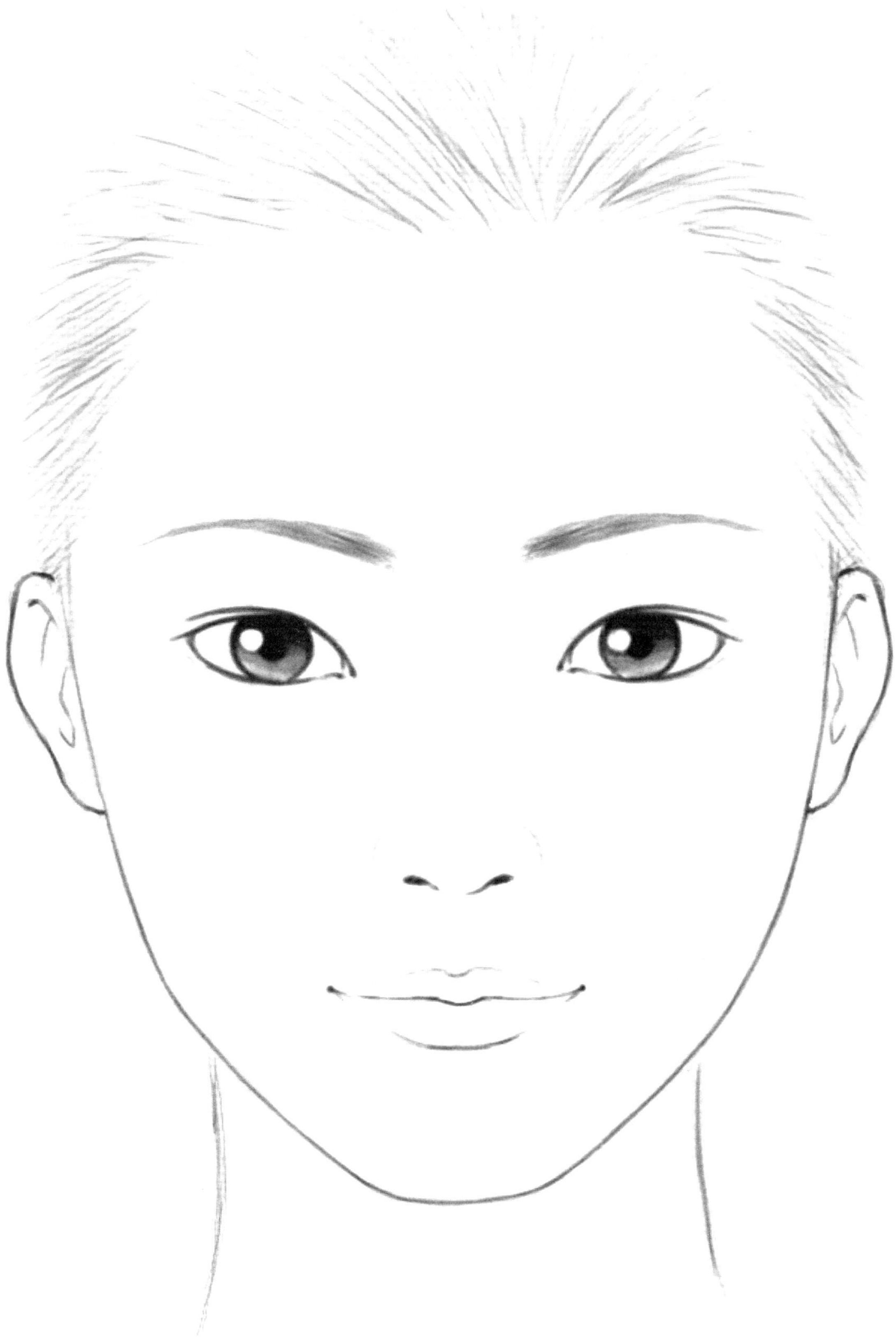

Make-up Title :		Skin Color :	
Eye-Shadow			
High-Light Color	Main Color	Point Color	Accent Color
Cheek Color		Lip Color	

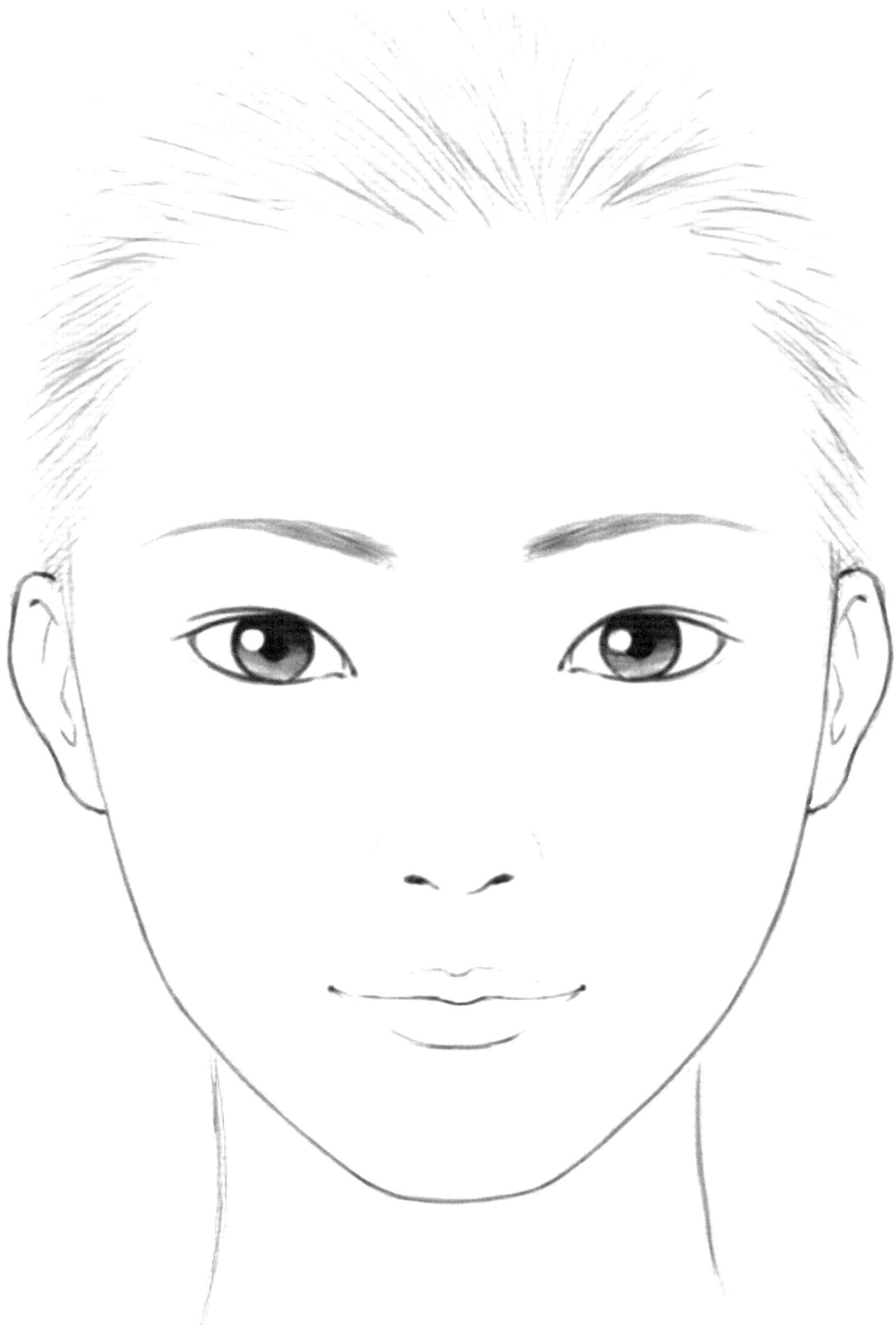

Make-up Title :		Skin Color :	
Eye-Shadow			
High-Light Color	Main Color	Point Color	Accent Color
Cheek Color		Lip Color	

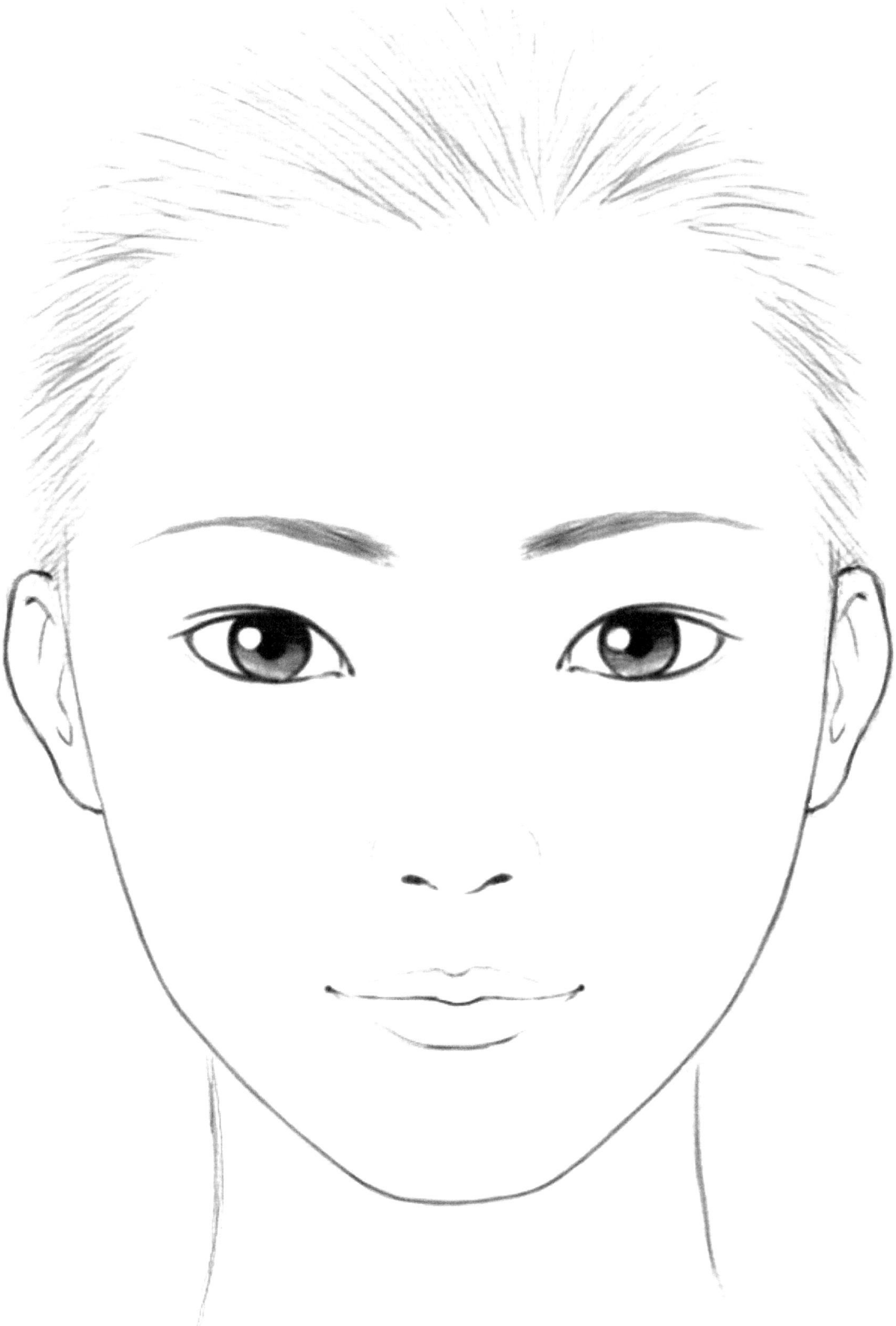

Make-up Title :		Skin Color :	
Eye-Shadow			
High-Light Color	Main Color	Point Color	Accent Color
Cheek Color		Lip Color	

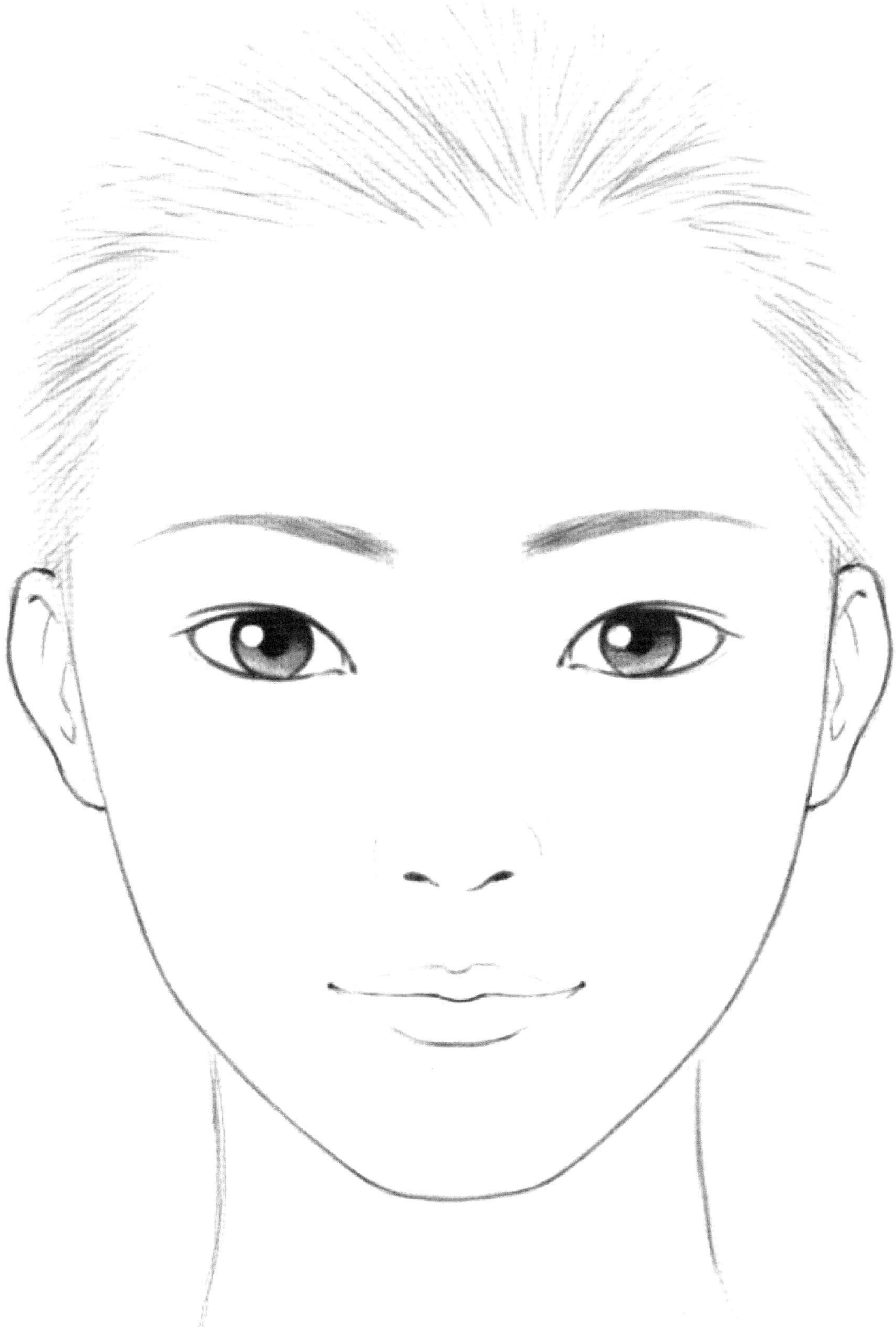

Make-up Title :		Skin Color :	
Eye-Shadow			
High-Light Color	Main Color	Foint Color	Accent Color
Cheek Color		Lip Color	

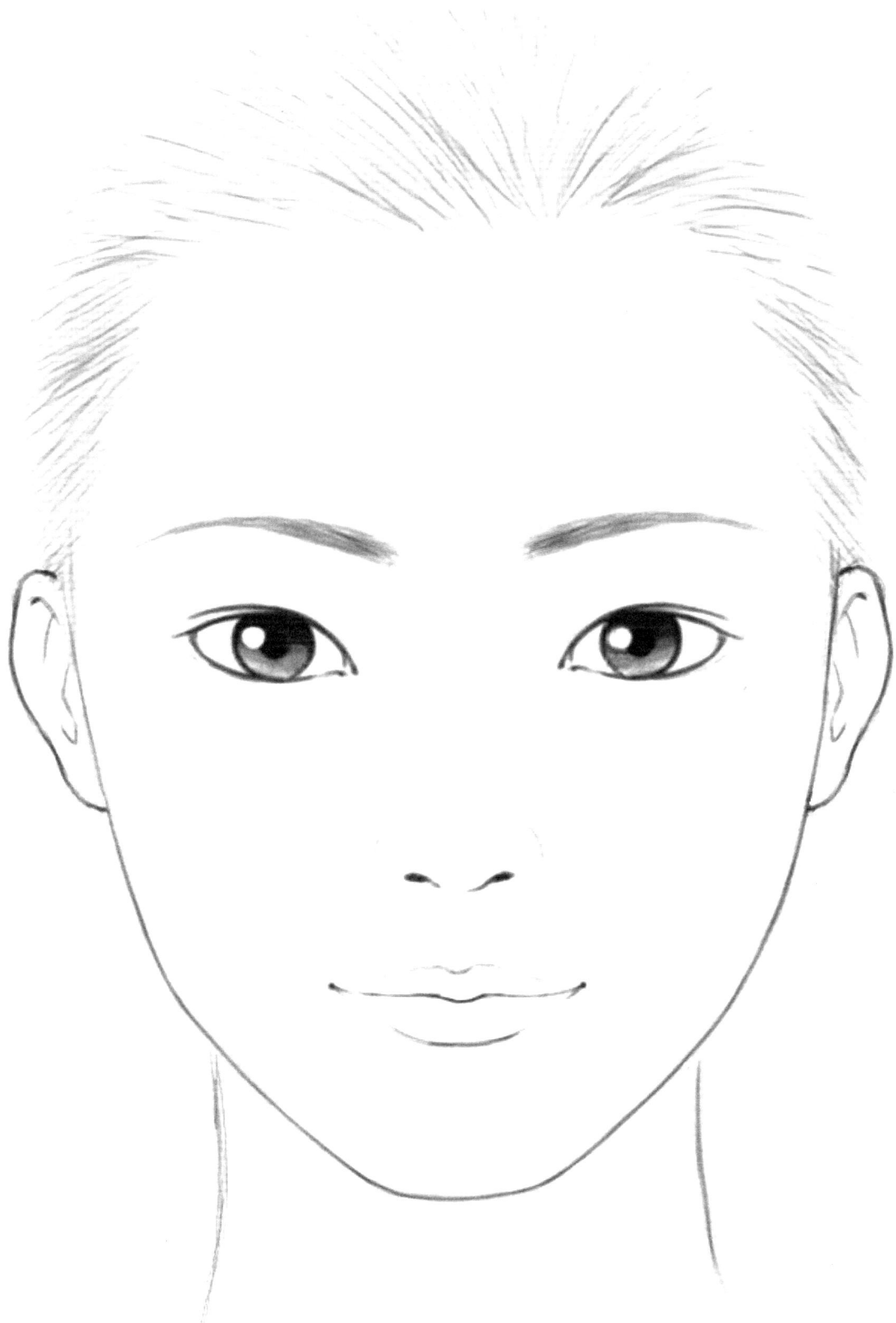

Make-up Title :		Skin Color :	
Eye-Shadow			
High-Light Color	Main Color	Point Color	Accent Color
Cheek Color		Lip Color	

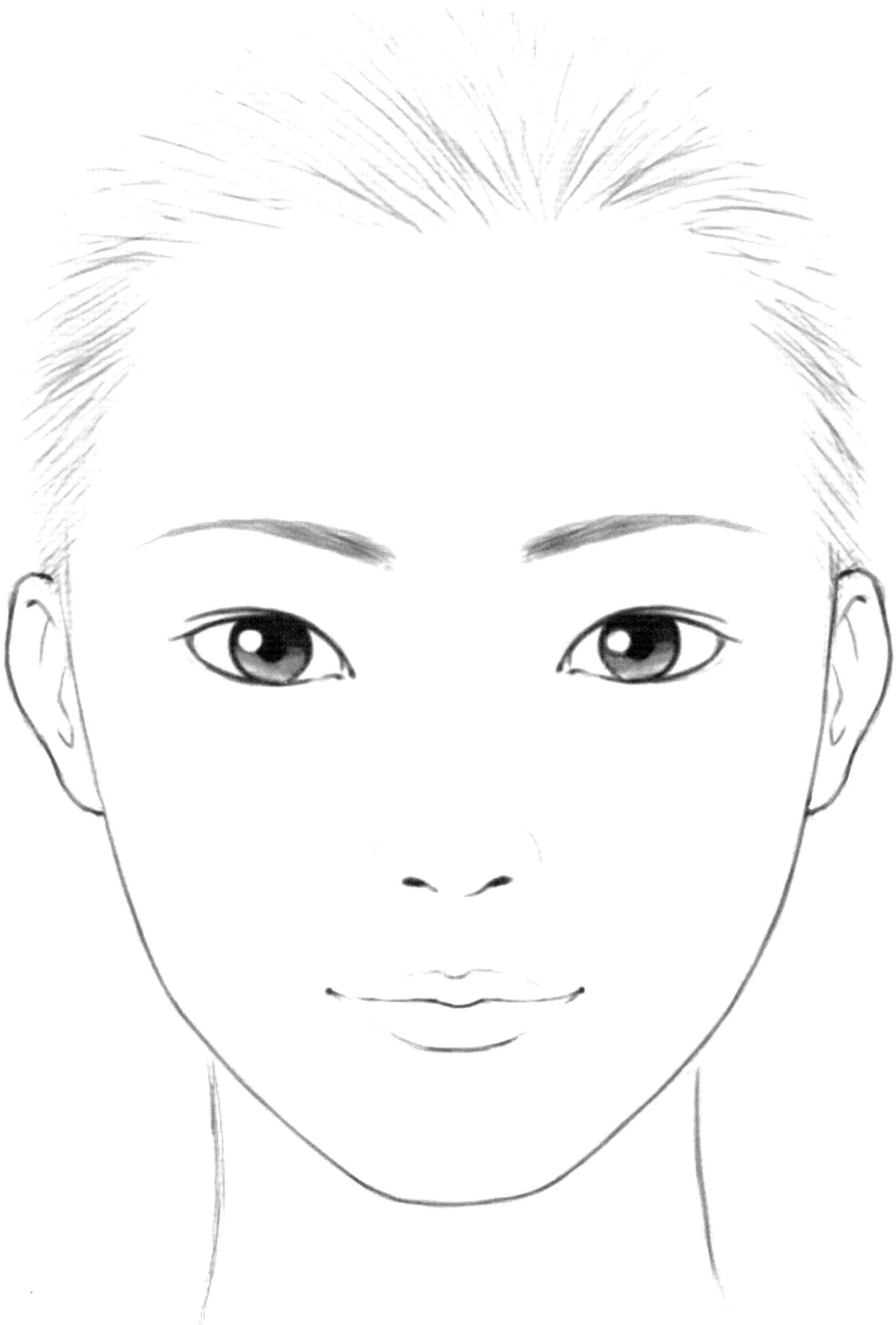

Make-up Title :		Skin Color :	
Eye-Shadow			
High-Light Color	Main Color	Point Color	Accent Color
Cheek Color		Lip Color	

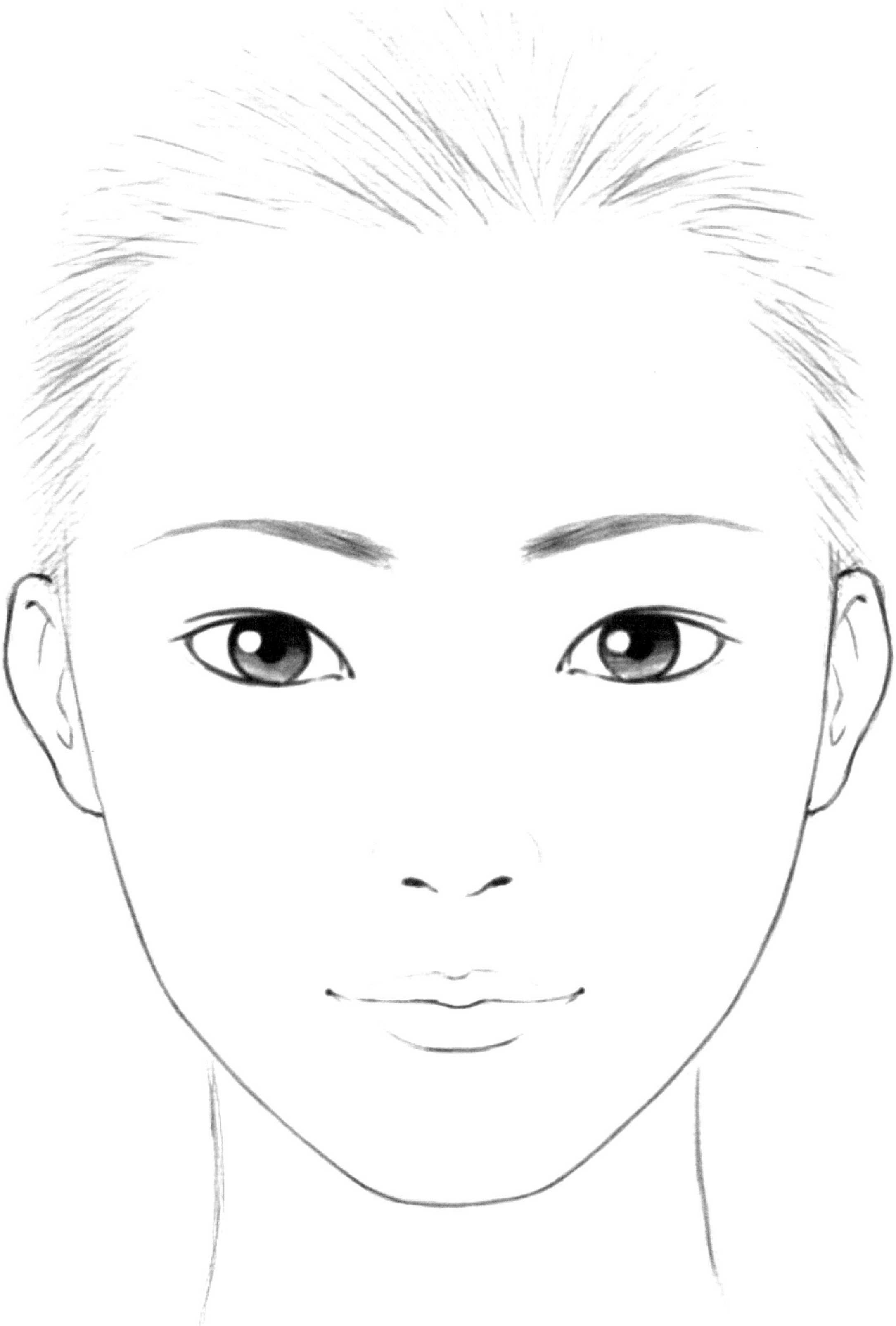

Make-up Title :		Skin Color :	
Eye-Shadow			
High-Light Color	Main Color	Point Color	Accent Color
Cheek Color		Lip Color	

Make-up Title :		Skin Color :	
Eye-Shadow			
High-Light Color	Main Color	Point Color	Accent Color
Cheek Color		Lip Color	

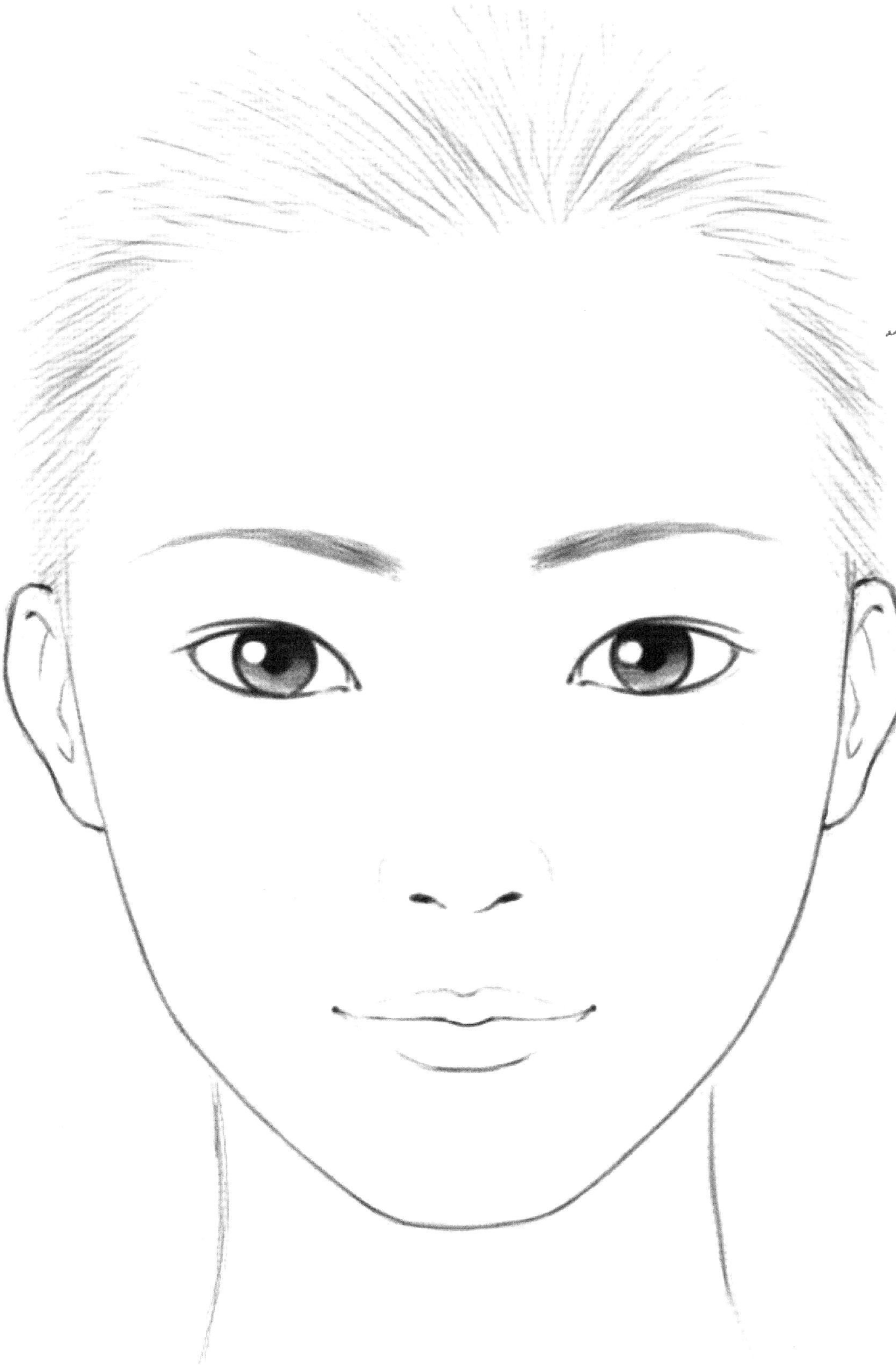

Make-up Title :		Skin Color :	
Eye-Shadow			
High-Light Color	Main Color	Point Color	Accent Color
Cheek Color		Lip Color	

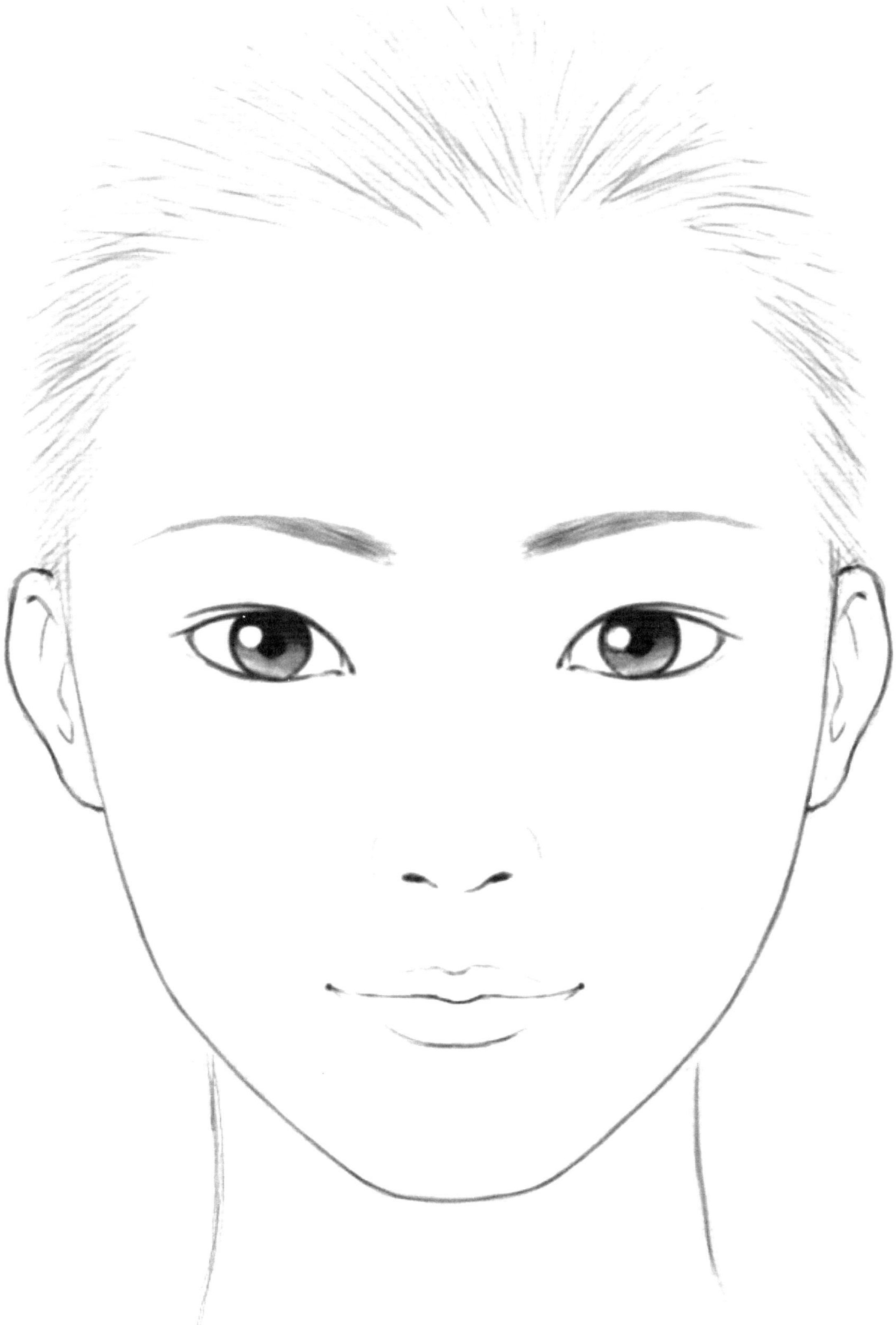

Make-up Title :		Skin Color :	
Eye-Shadow			
High-Light Color	Main Color	Point Color	Accent Color
Cheek Color		Lip Color	

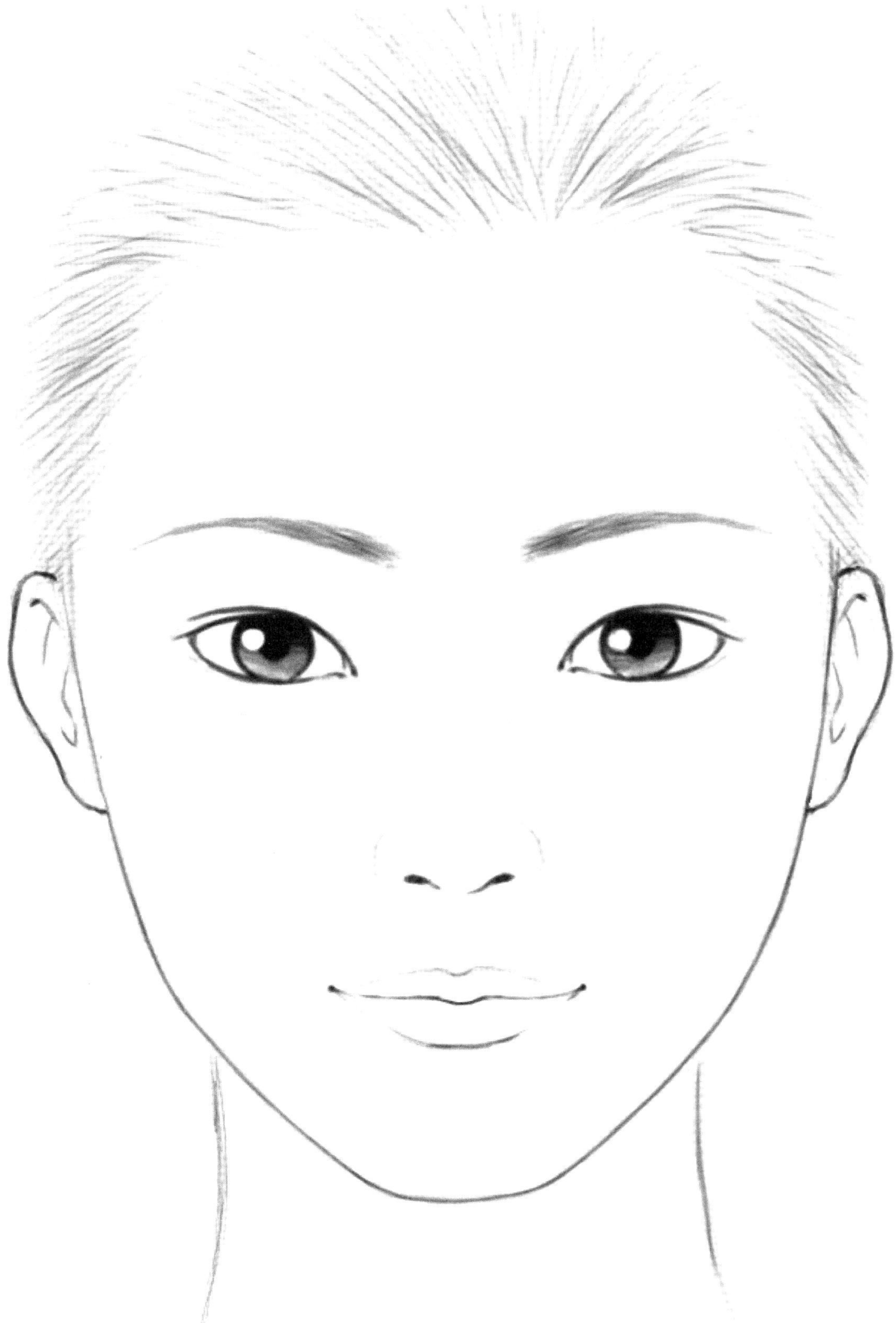

Make-up Title :		Skin Color :	
Eye-Shadow			
High-Light Color	Main Color	Point Color	Accent Color
Cheek Color		Lip Color	

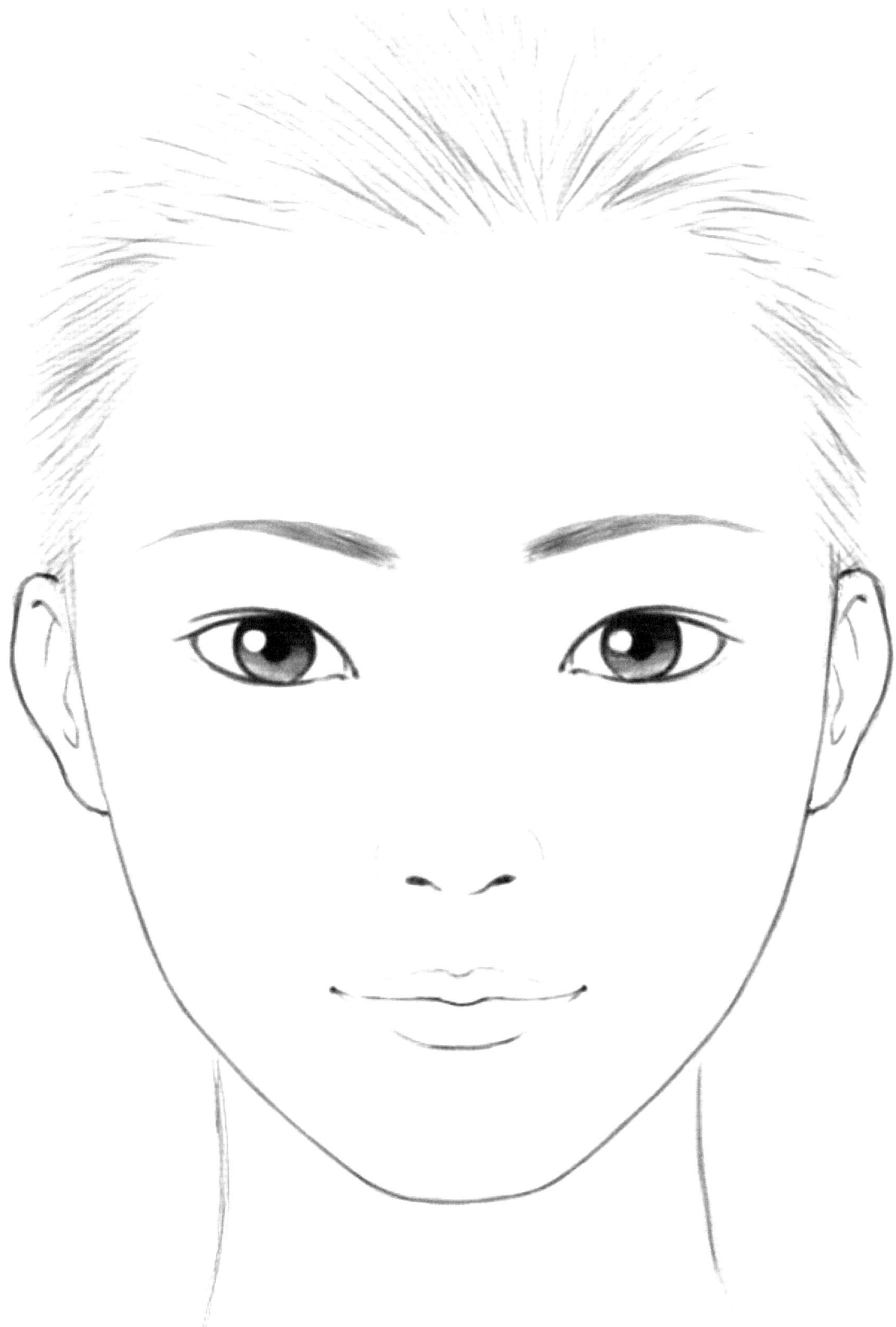

Make-up Title :		Skin Color :	
Eye-Shadow			
High-Light Color	Main Color	Point Color	Accent Color
Cheek Color		Lip Color	

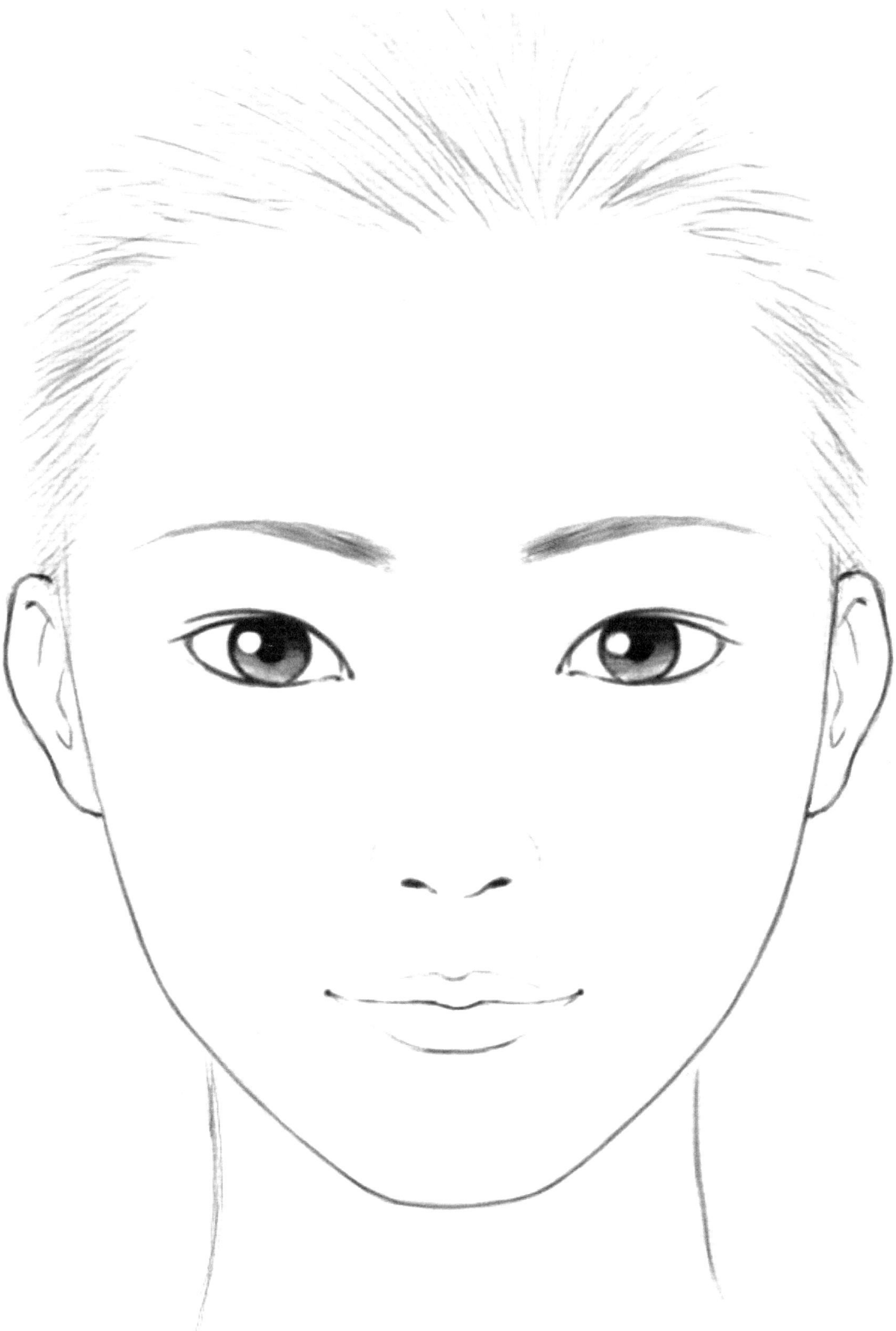

Make-up Title :		Skin Color :	
Eye-Shadow			
High-Light Color	Main Color	Point Color	Accent Color
Cheek Color		Lip Color	

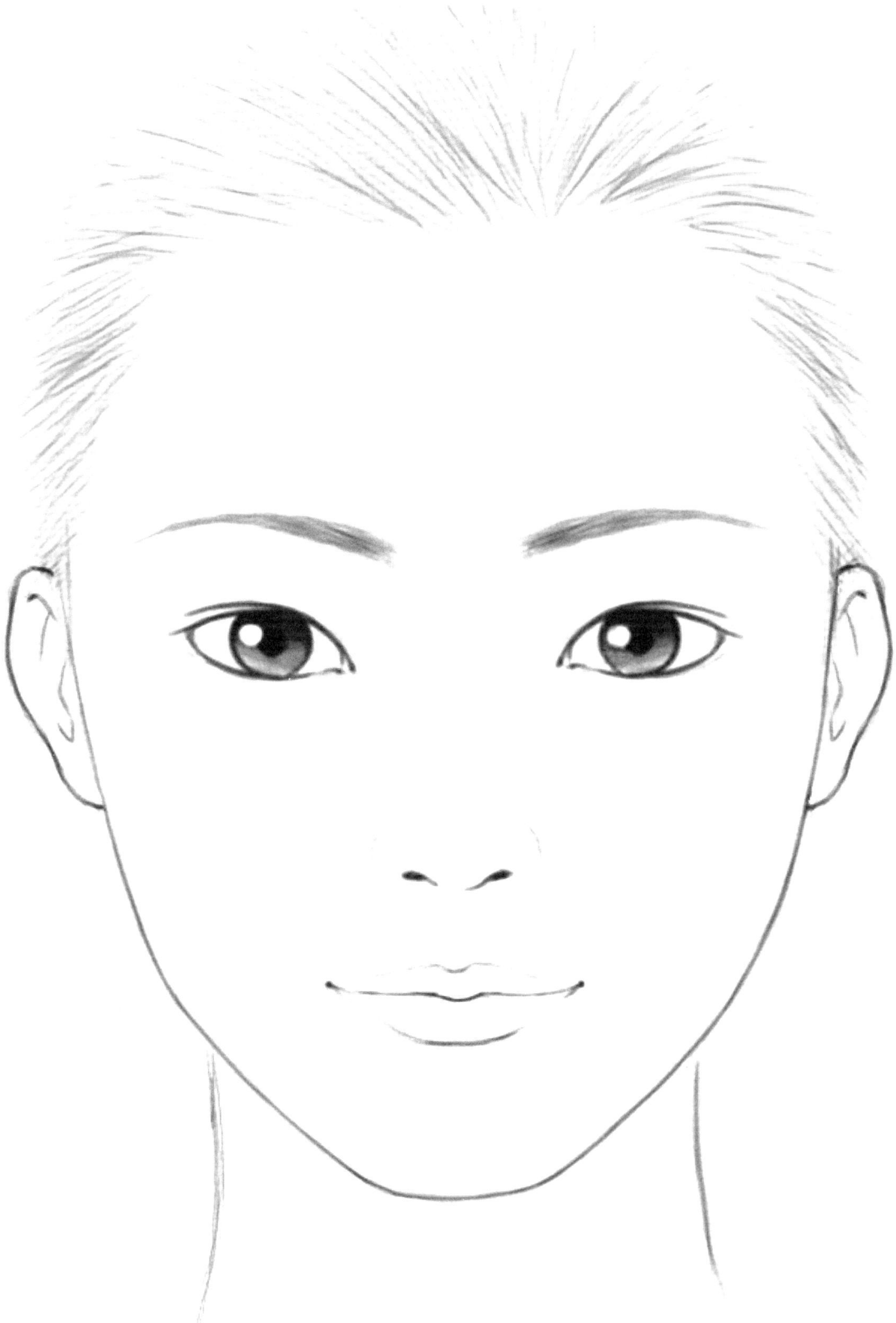

Make-up Title :		Skin Color :	
Eye-Shadow			
High-Light Color	Main Color	Point Color	Accent Color
Cheek Color		Lip Color	

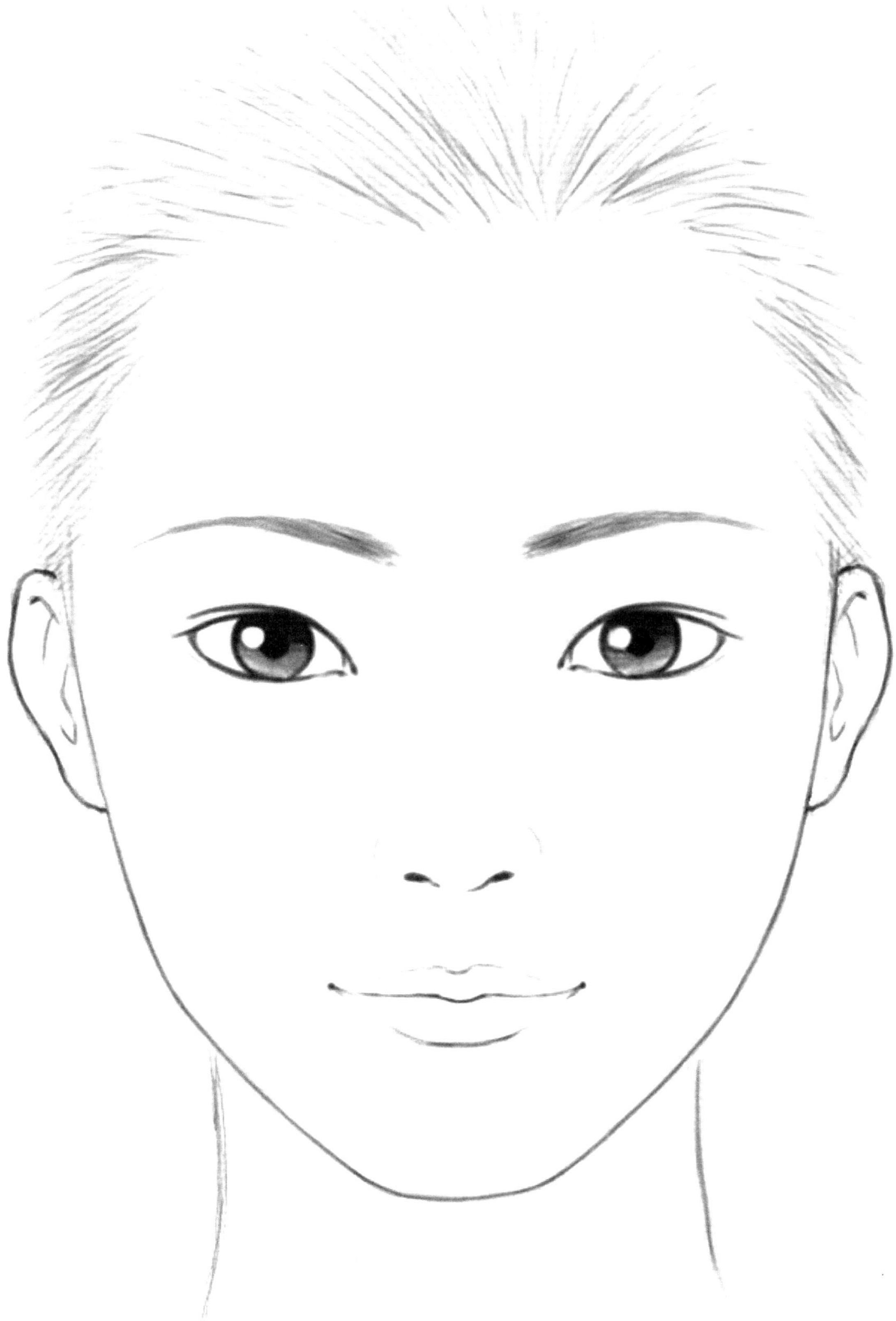

Make-up Title :		Skin Color :	
Eye-Shadow			
High-Light Color	Main Color	Point Color	Accent Color
Cheek Color		Lip Color	

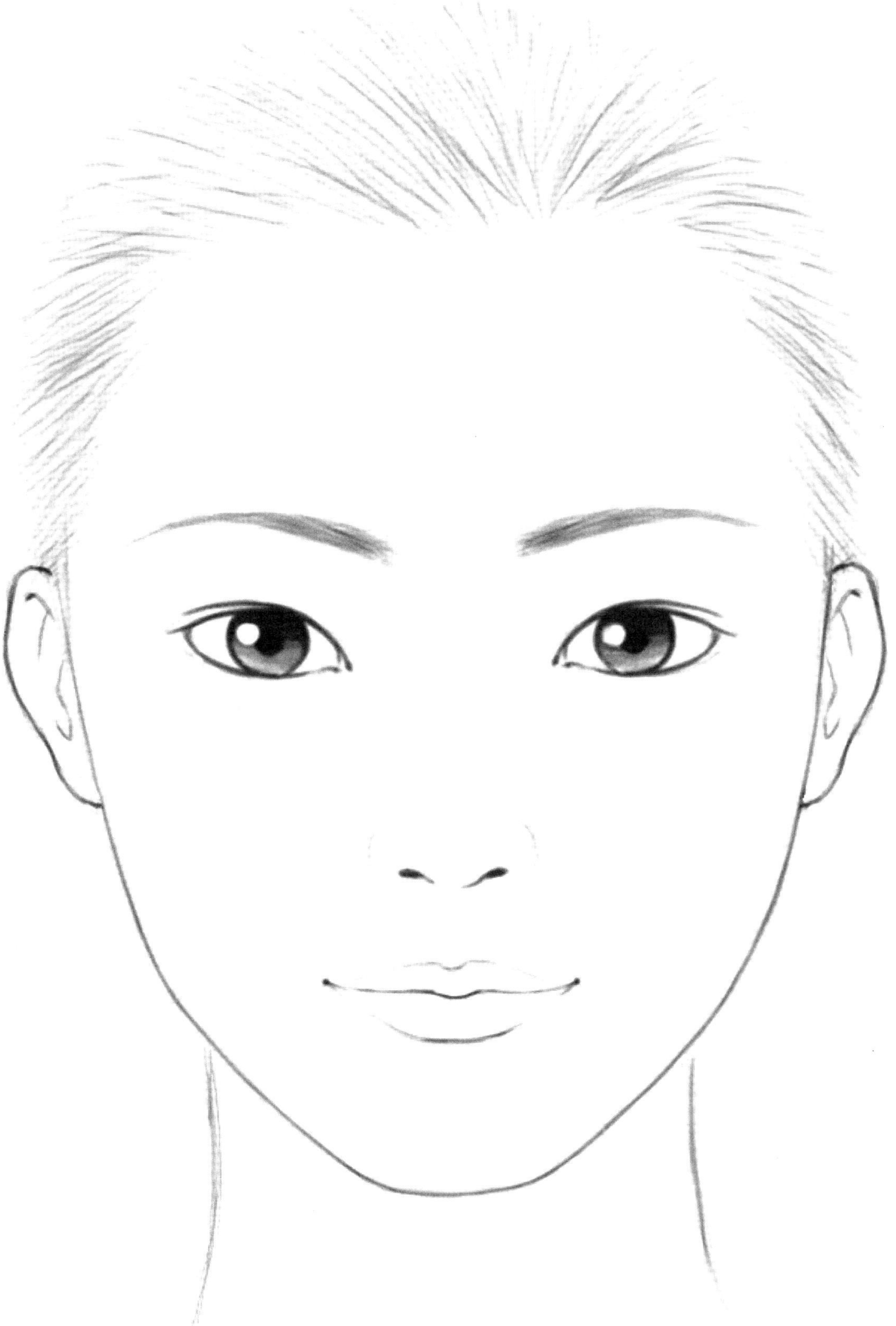

Make-up Title :		Skin Color :	
Eye-Shadow			
High-Light Color	Main Color	Point Color	Accent Color
Cheek Color		Lip Color	

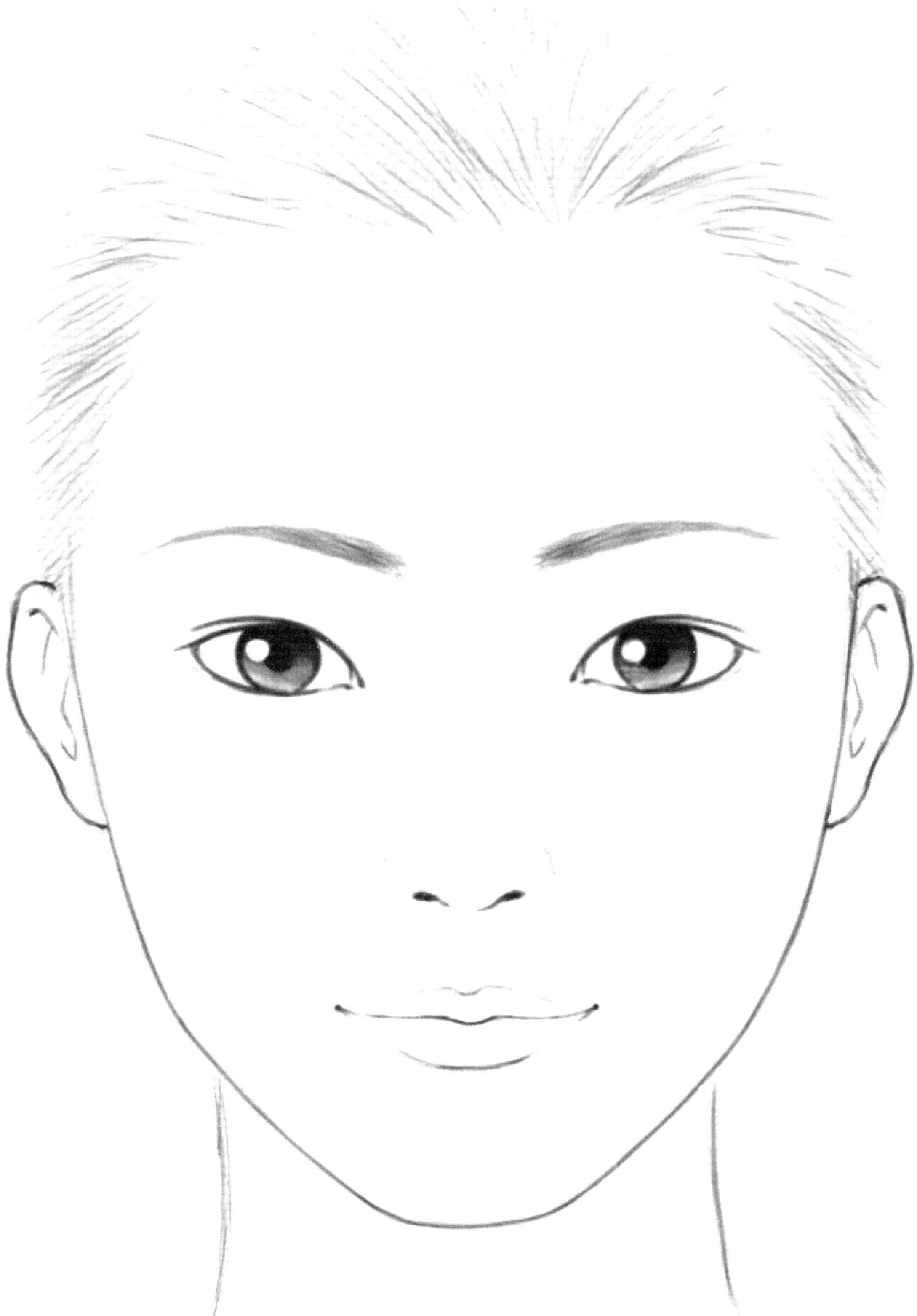

Make-up Title :		Skin Color :	
Eye-Shadow			
High-Light Color	Main Color	Point Color	Accent Color
Cheek Color		Lip Color	

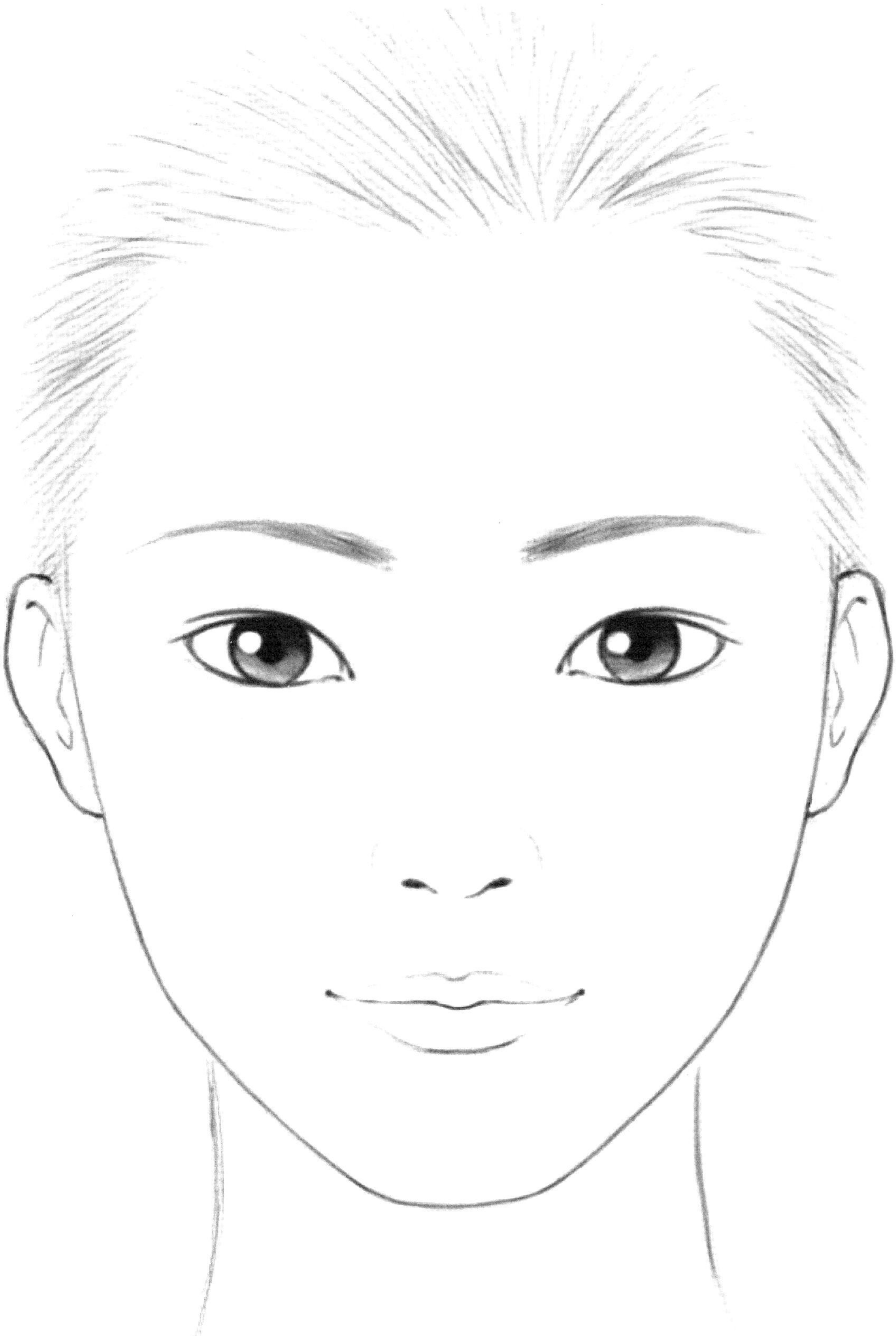

Make-up Title :		Skin Color :	
Eye-Shadow			
High-Light Color	Main Color	Point Color	Accent Color
Cheek Color		Lip Color	

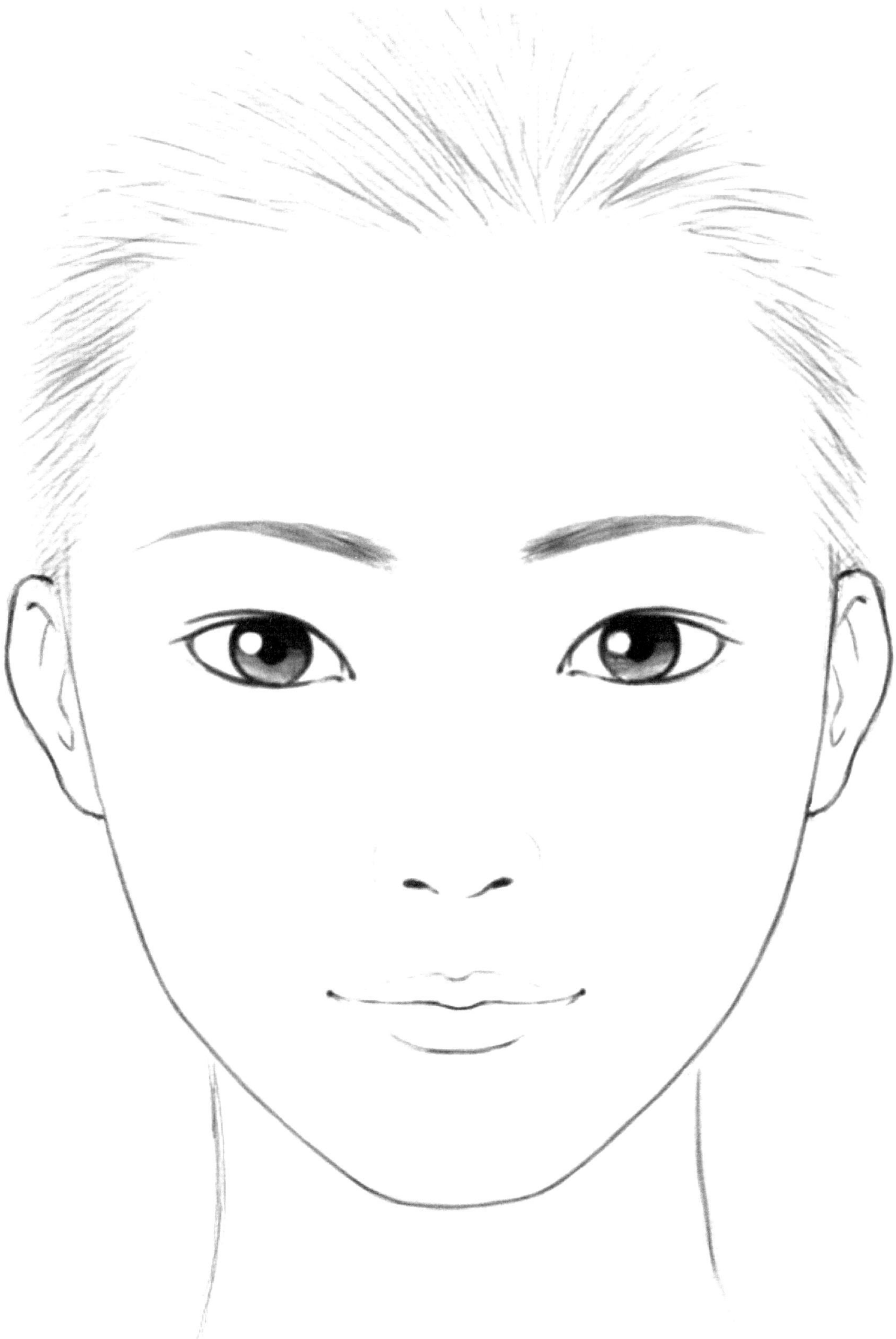

Make-up Title :		Skin Color :	
Eye-Shadow			
High-Light Color	Main Color	Point Color	Accent Color
Cheek Color		Lip Color	

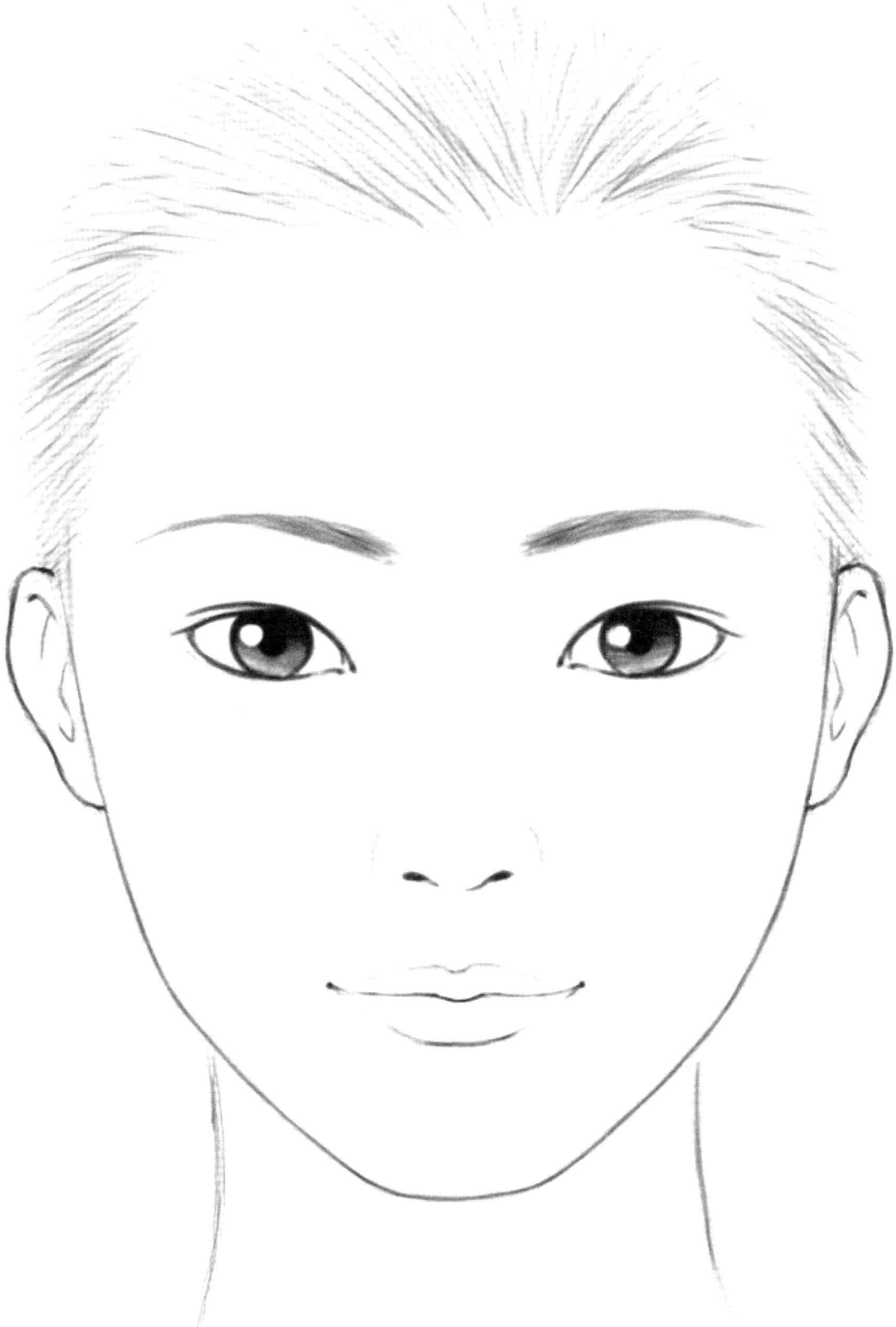

Make-up Title :		Skin Color :	
Eye-Shadow			
High-Light Color	Main Color	Point Color	Accent Color
Cheek Color		Lip Color	

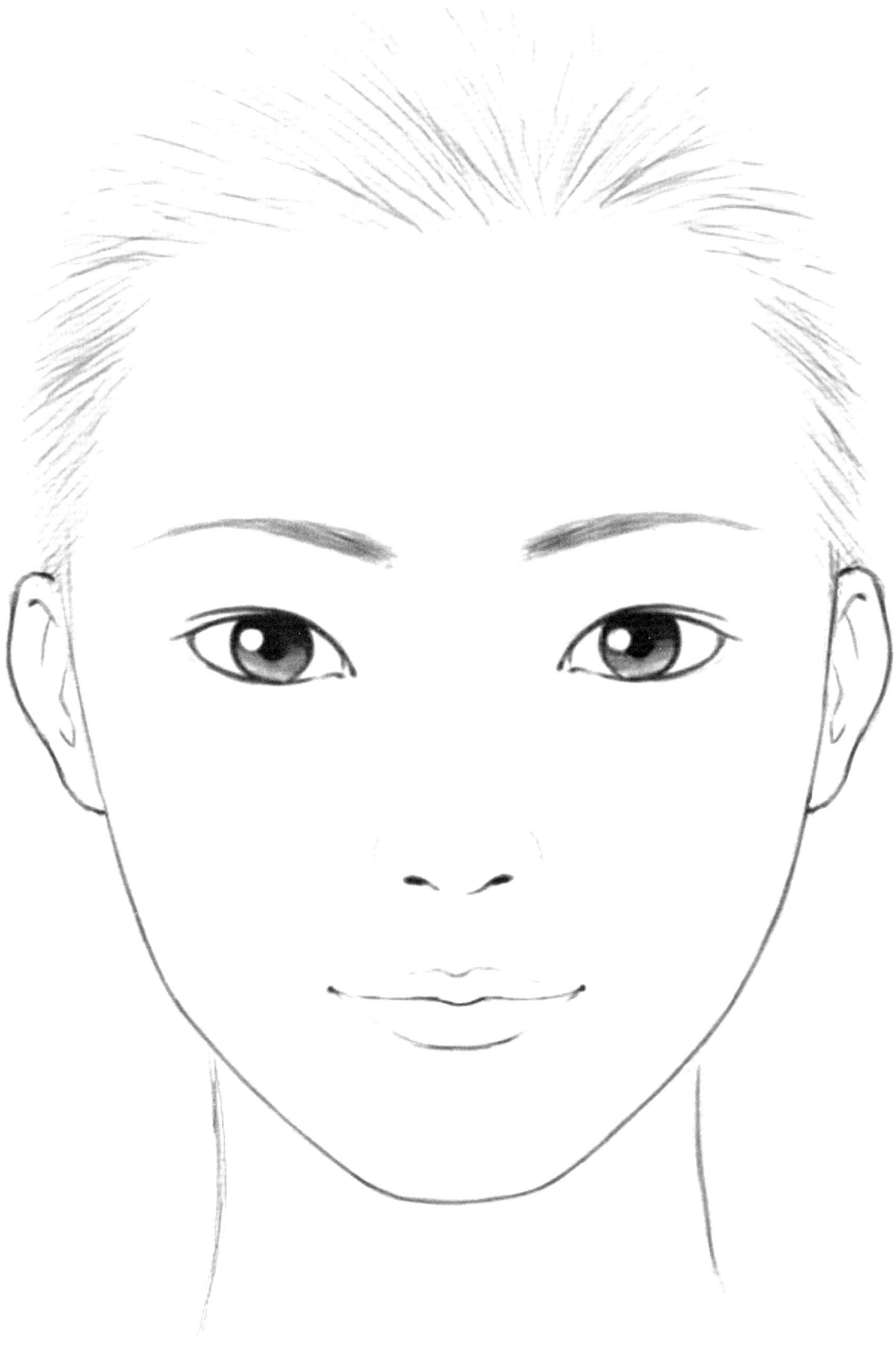

Make-up Title :		Skin Color :	
Eye-Shadow			
High-Light Color	Main Color	Point Color	Accent Color
Cheek Color		Lip Color	

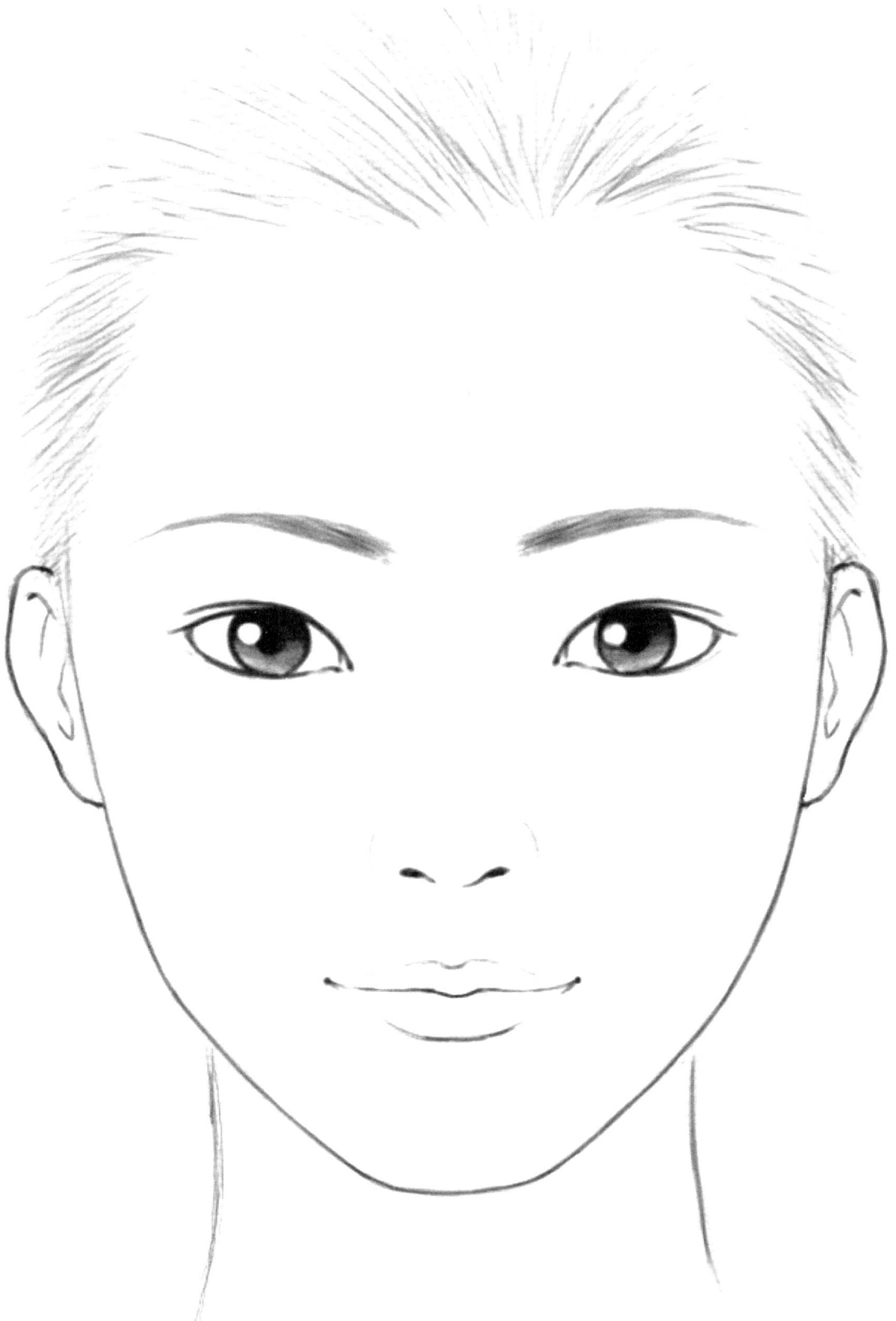

Make-up Title :		Skin Color :	
Eye-Shadow			
High-Light Color	Main Color	Point Color	Accent Color
Cheek Color		Lip Color	

Chapter 12

남자 얼굴 정면

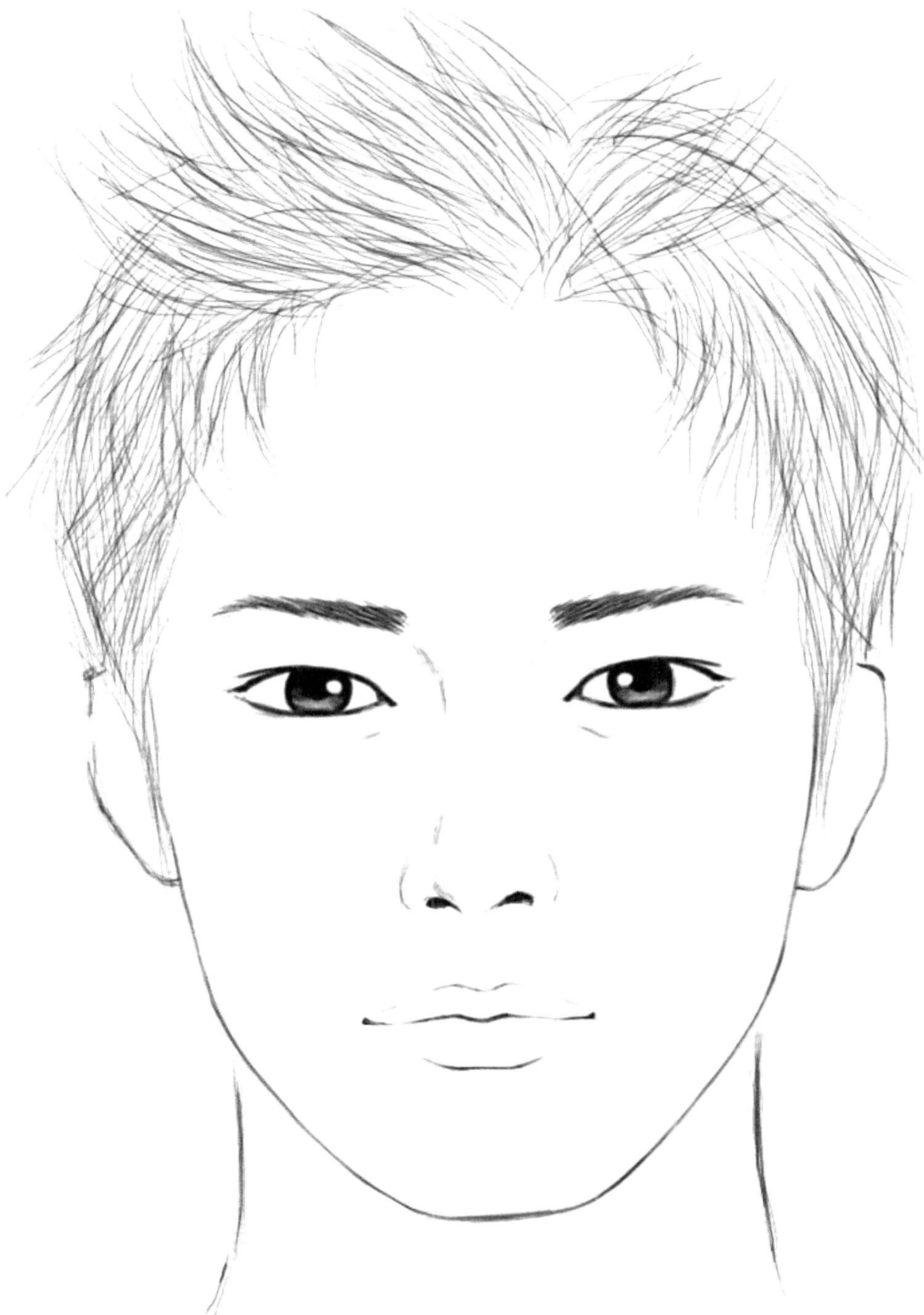

Make-up Title :		Skin Color :	
Eye-Shadow			
High-Light Color	Main Color	Point Color	Accent Color
Cheek Color		Lip Color	

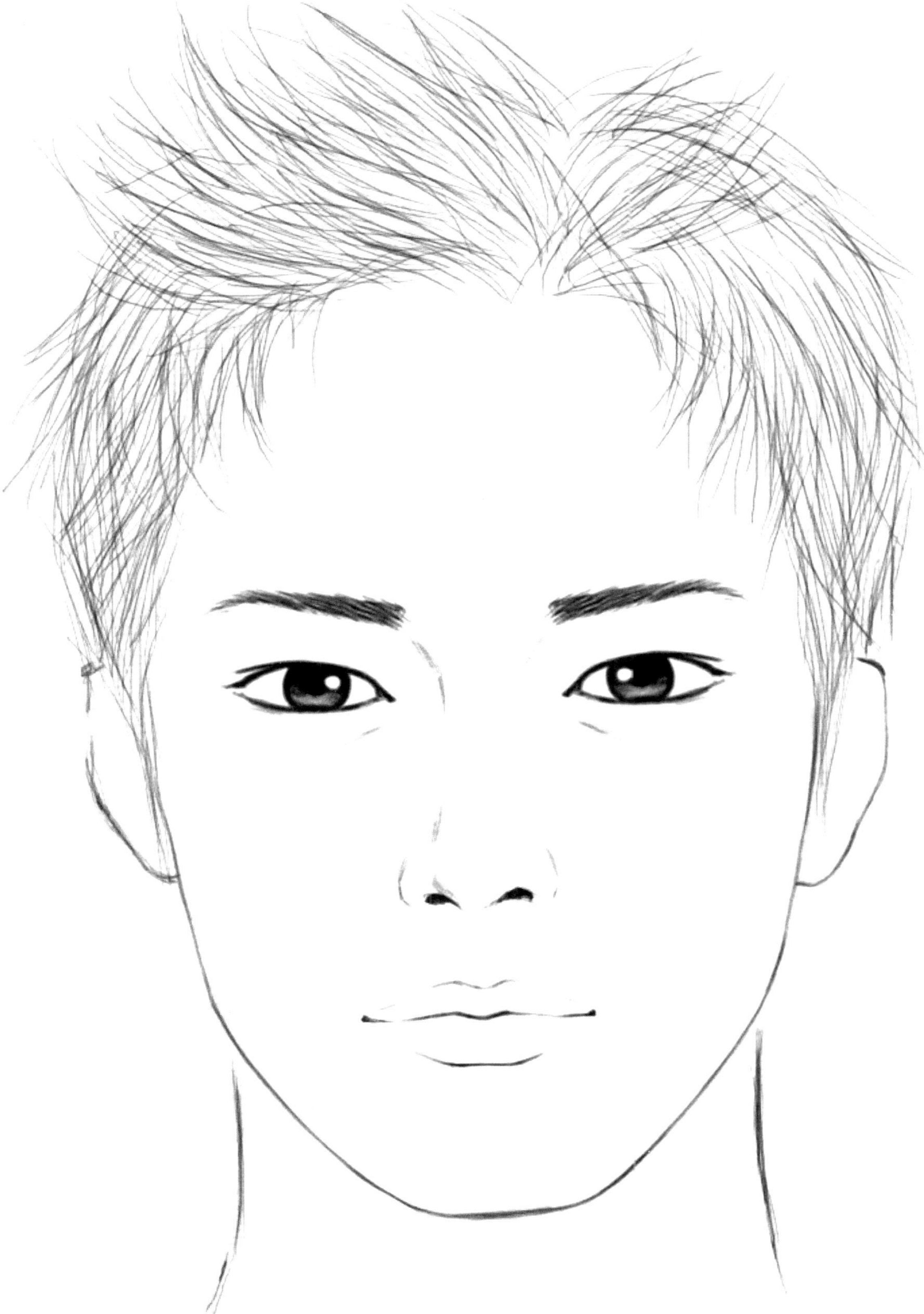

Make-up Title :		Skin Color :	
Eye-Shadow			
High-Light Color	Main Color	Point Color	Accent Color
Cheek Color		Lip Color	

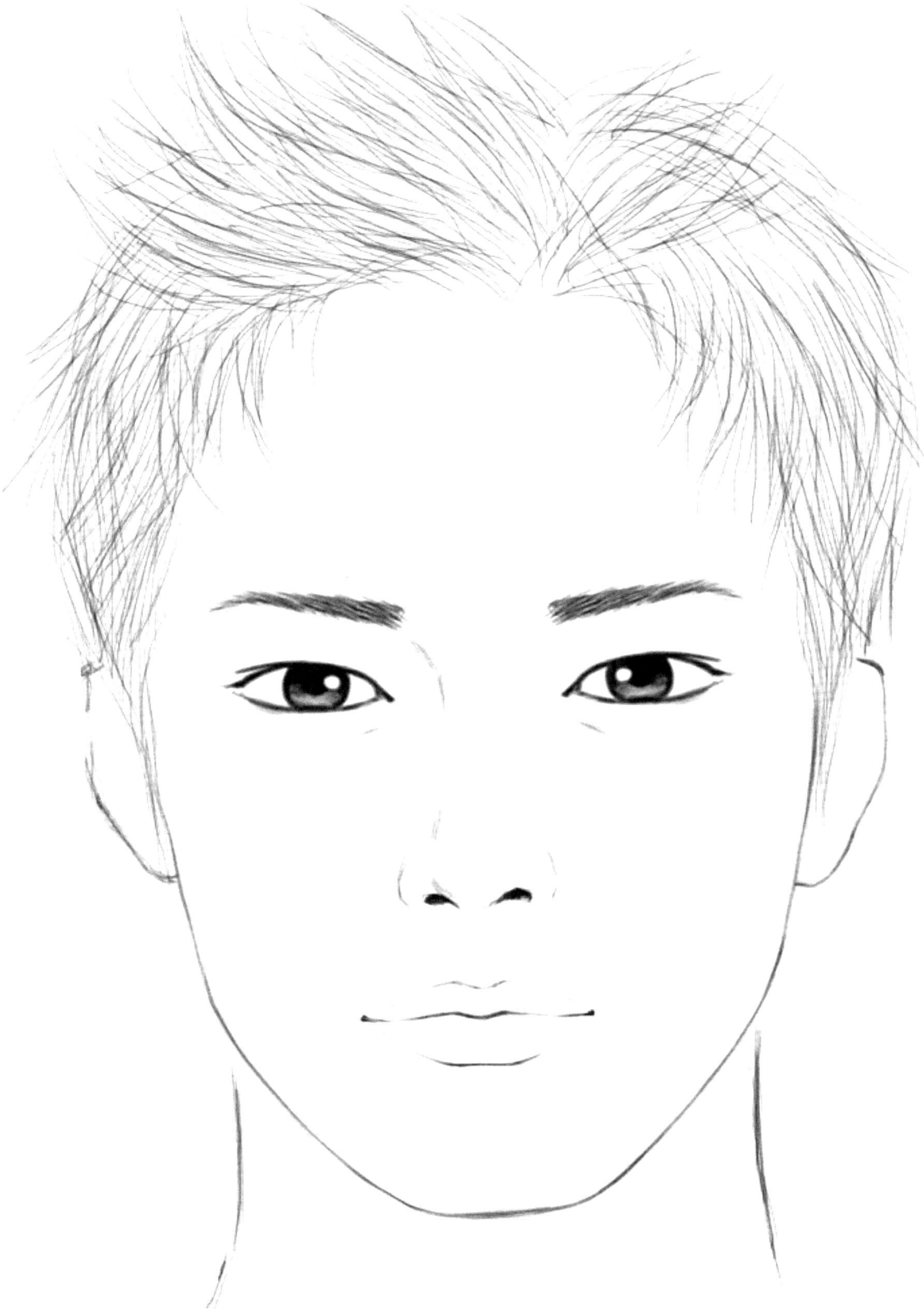

Make-up Title :		Skin Color :	
Eye-Shadow			
High-Light Color	Main Color	Point Color	Accent Color
Cheek Color		Lip Color	

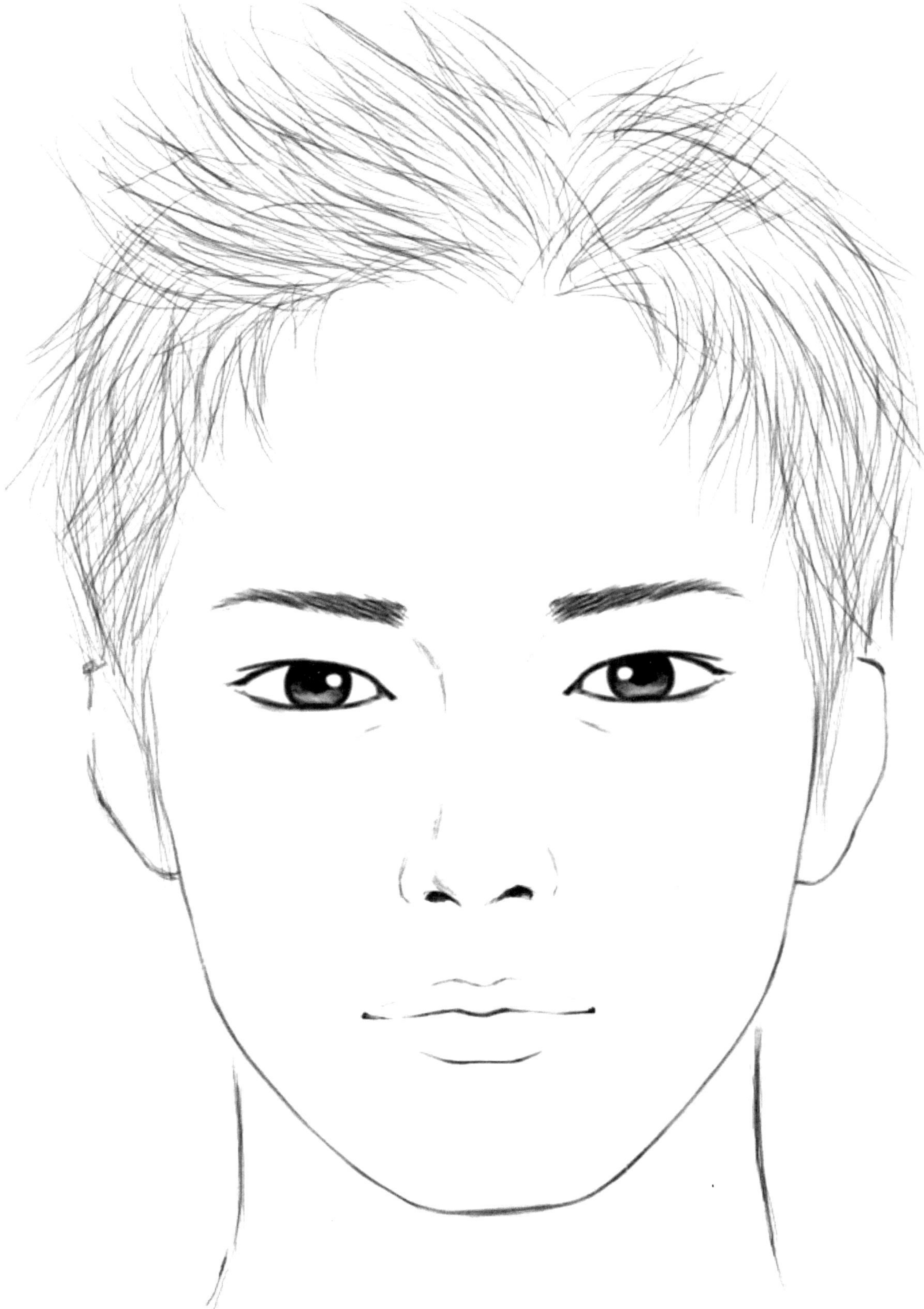

Make-up Title :		Skin Color :	
Eye-Shadow			
High-Light Color	Main Color	Point Color	Accent Color
Cheek Color		Lip Color	

Chapter 13

여자 상반신

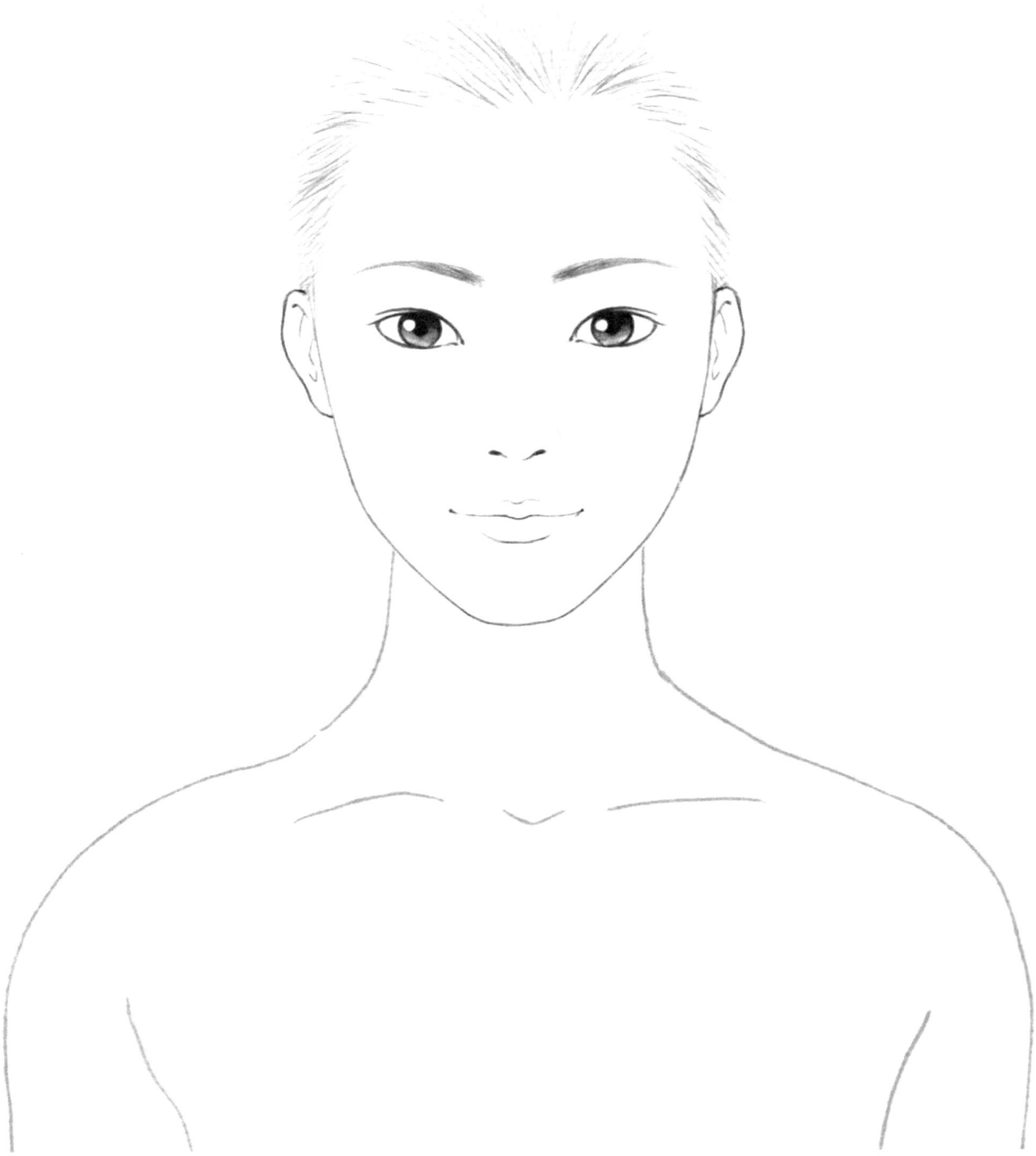

Make-up Title :		Skin Color :	
Eye-Shadow			
High-Light Color	Main Color	Point Color	Accent Color
Cheek Color		Lip Color	

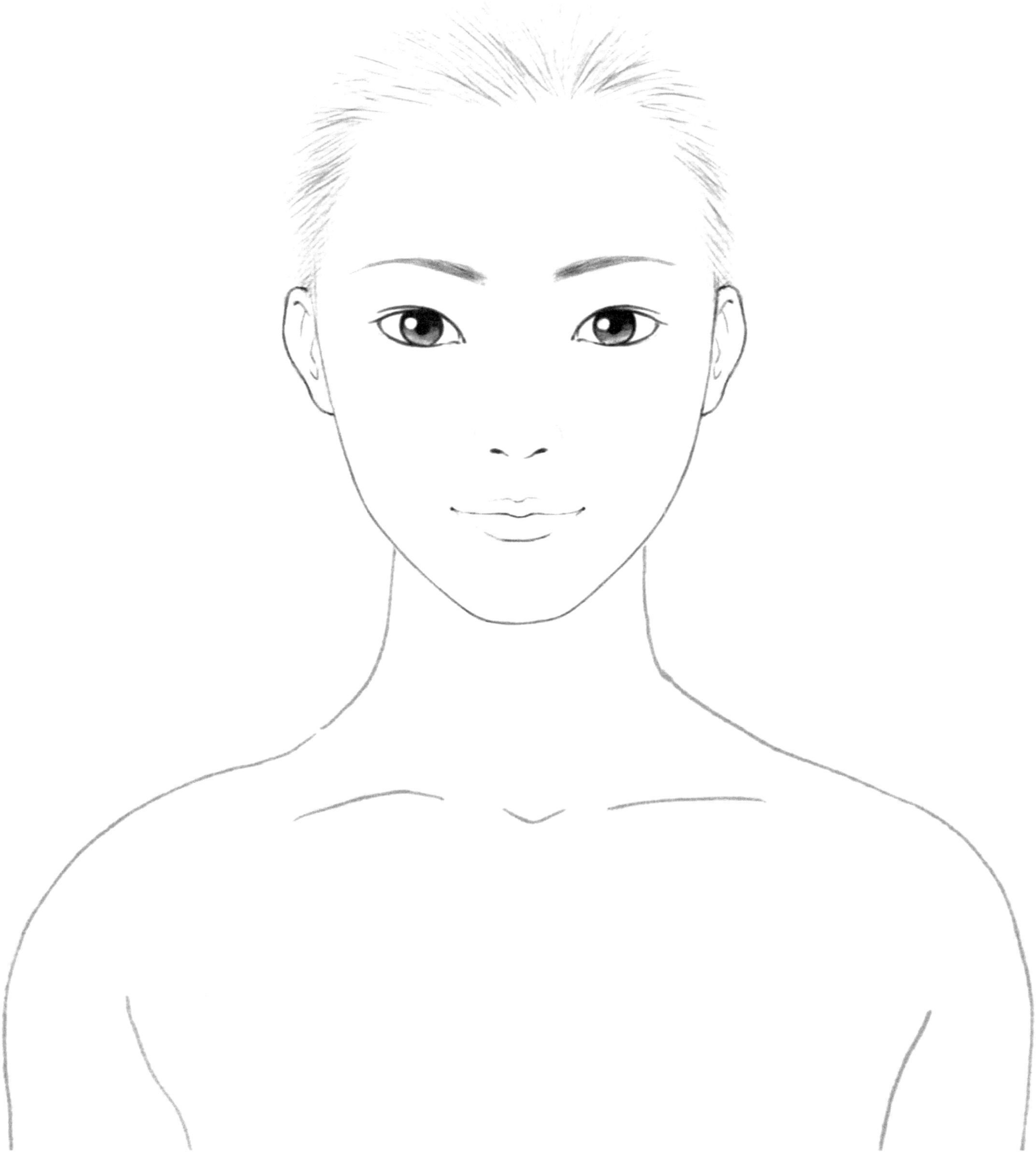

Make-up Title :		Skin Color :	
Eye-Shadow			
High-Light Color	Main Color	Point Color	Accent Color
Cheek Color		Lip Color	

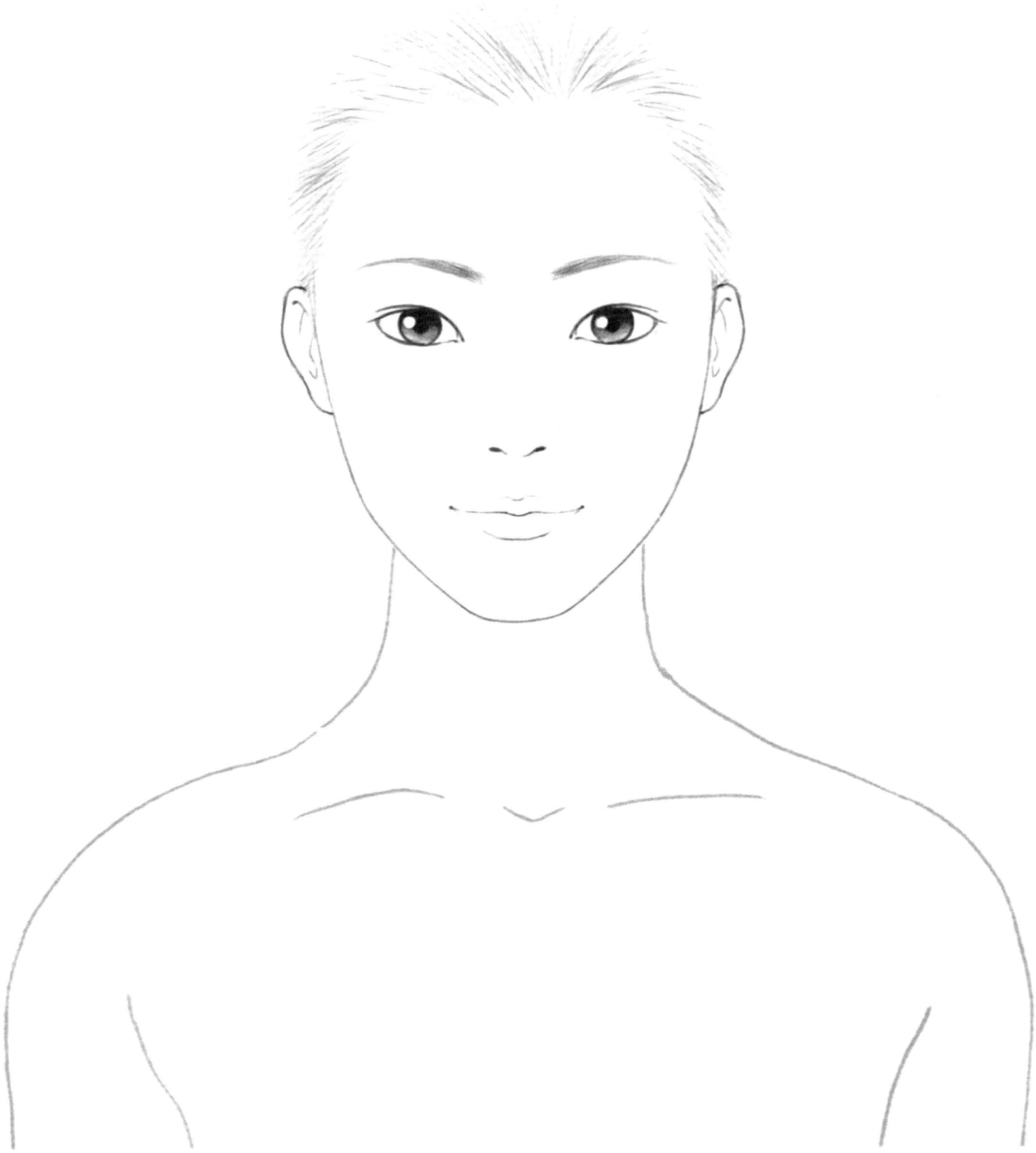

Make-up Title :		Skin Color :	
Eye-Shadow			
High-Light Color	Main Color	Point Color	Accent Color
Cheek Color		Lip Color	

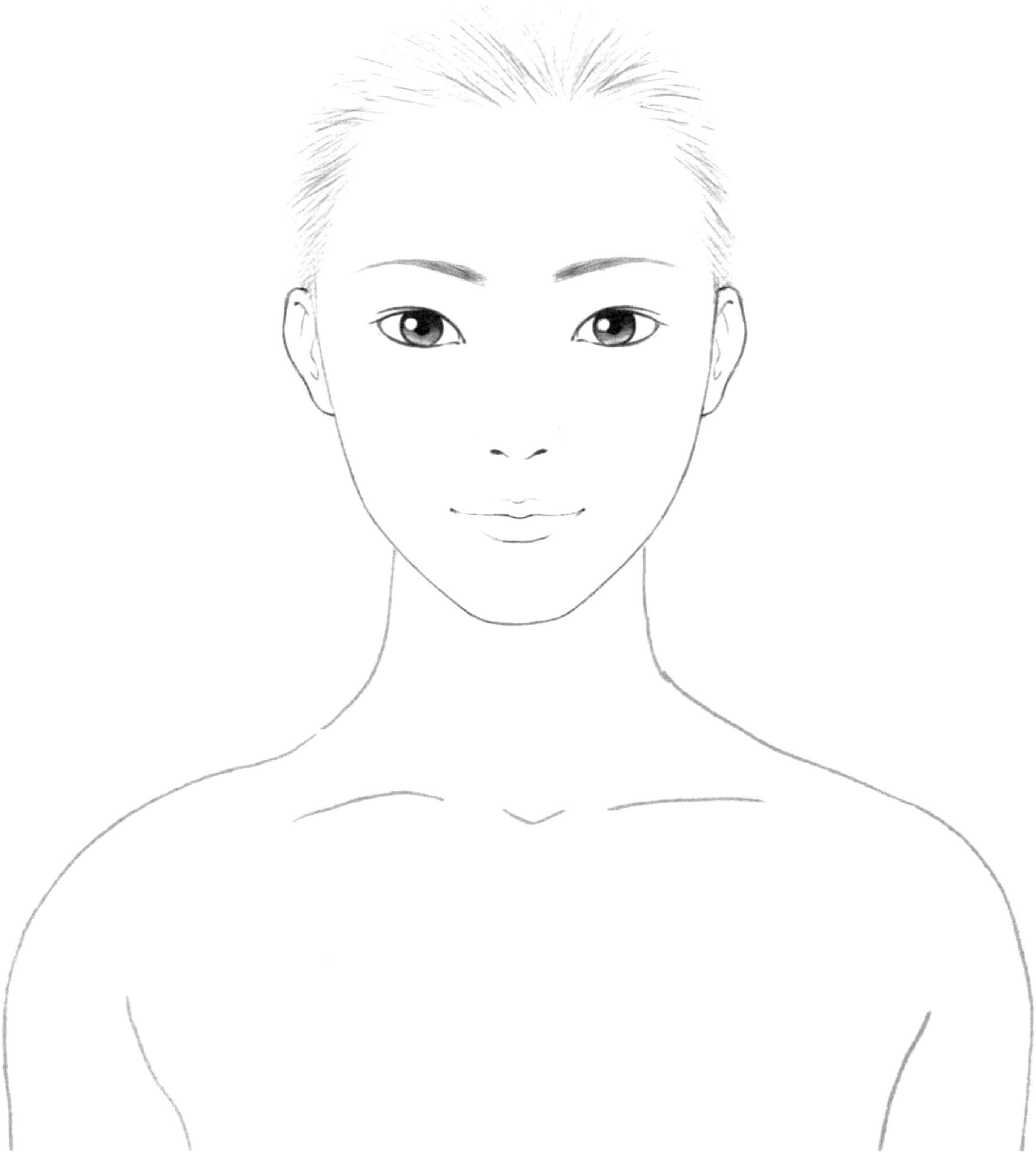

Make-up Title :		Skin Color :	
Eye-Shadow			
High-Light Color	Main Color	Point Color	Accent Color
Cheek Color		Lip Color	

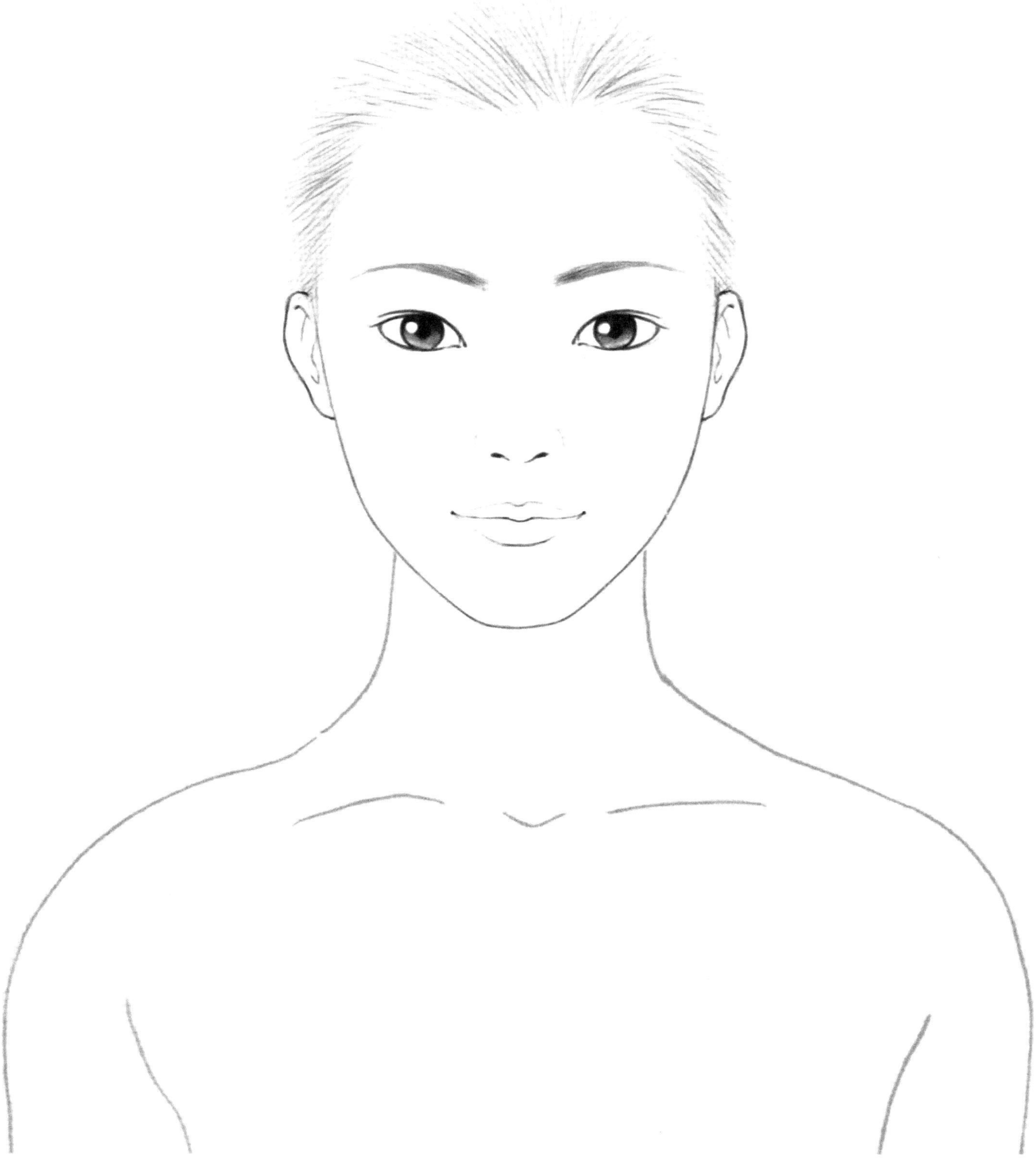

Make-up Title :		Skin Color :	
Eye-Shadow			
High-Light Color	Main Color	Point Color	Accent Color
Cheek Color		Lip Color	

Chapter 14

메이크업 국가 자격증 실기 패턴과 구성

미용사(메이크업) 기능사 과제 유형

과제 유형	1과제(40분)	2과제(40분)	3과제(50분)	4과제(25분)
	뷰티 메이크업	시대 메이크업	캐릭터 메이크업	속눈썹 익스텐션 및 수염
작업대상	모델			마네킹
세부과제	웨딩(로맨틱)	1930년대 그레타 가르보	레오파드 이미지	왼쪽 속눈썹 익스텐션
	웨딩(클래식)	1950년대 마릴린 먼로	무용(고전)	오른쪽 속눈썹 익스텐션
	한복	1960년대 트위기	무용(발레)	미디어 수염
	내추럴	1970~1980 펑크	노역	
배점	30	30	25	15

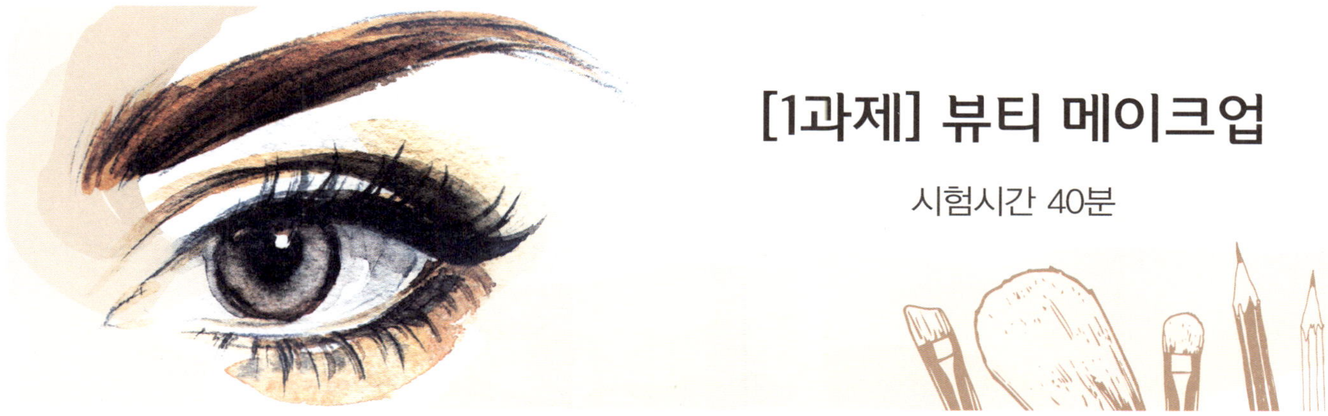

[1과제] 뷰티 메이크업

시험시간 40분

Point

로맨틱 웨딩 : 한 톤 밝은 베이스
둥근 아이브로우
핑크색 아이 쉐도우
둥근 cheek
립그로스 마무리

클래식 웨딩 : 매트한 피부
각잔 눈썹
피치 아이 쉐도우
사선으로 cheek
베이지 핑크 립 스틱

한복 : 갈색 아이브로우
브라운 피치 아이쉐도우
애굣살
사선형태의 오렌지 cheek

네추럴 : 자연스러운 속눈썹
피치색 치크
베이지 핑크립

Make-up Title : 웨딩로맨틱		Skin Color : 피부보다 한 톤 밝게 표현	
Eye-Shadow			
High-Light Color	Main Color	Point Color	Accent Color
White	Peal Pink	Peal Violet	
Cheek Color		Lip Color	
Pink(애플존)		Pink	

Make-up Title : 웨딩로맨틱		Skin Color : Foundation은 모델의 피부색에 맞춰 결점 커버	
Eye-Shadow			
High-Light Color	Main Color	Point Color	Accent Color
Gold Peal	Peach	Brown	
Cheek Color		Lip Color	
Peach(사선으로)		Beige Pink	

Make-up Title : 한복 메이크업		Skin Color : 피부색과 비슷하게 표현	
Eye-Shadow			
High-Light Color	Main Color	Point Color	Accent Color
Cream	Peal Peach	Brown	
Cheek Color		Lip Color	
Orange		Orange Red	

Make-up Title : 네츄럴 메이크업		Skin Color : 피부색에 맞는 Liquid Foundation으로 결점 커버	
Eye-Shadow			
High-Light Color	Main Color	Point Color	Accent Color
White	Beige	Brown	
Cheek Color		Lip Color	
Peach		Beige Pink	

[2과제] 시대 메이크업

시험시간 40분

Point

그레타 가르보 : 아치형의 가는 아이 브로우
아이홀 강조
광대뼈에 핑크색 치크
광대뼈 아래 브라운 쉐딩

마릴린 먼로 : 밝은 핑크톤의 매트한 표현
각진 아브로우
긴 아이라이너
애굣점

트위기 : 갈색의 각진 눈썹
화이트, 핑크,네이비, 그레이, 청색의 인위적 쌍겹
언더 속눈썹 강조
베이지 핑크 립

펑크 : 창백한 피부
눈썹 강조,
아이홀 강조
홀라인에 검은 아이라이너,
레드 브라운 치크,
다크 레드 립

Make-up Title : 그레타 가르보		Skin Color : 피부톤과 비슷하게 표현	
Eye-Shadow			
High-Light Color	Main Color	Point Color	Accent Color
White	Beige	Brown	
Cheek Color		Lip Color	
Pink		Red	

Make-up Title : 마릴린 먼로		Skin Color : 밝은 핑크톤	
Eye-Shadow			
High-Light Color	Main Color	Point Color	Accent Color
White	Beige Pink	L Light Brown	
Cheek Color		Lip Color	
Pink		Red	

Make-up Title : 트위기		Skin Color : 모델의 피부 톤	
Eye-Shadow			
High-Light Color	Main Color	Point Color	Accent Color
White	Pink	Gray	Navy
Cheek Color		Lip Color	
Pink(애플존)		Beige Pink	

Make-up Title : 펑크		Skin Color : 창백하게 표현	
Eye-Shadow			
High-Light Color	Main Color	Point Color	Accent Color
White	Beige		Black
Cheek Color		Lip Color	
Red Brown		Dark Red	

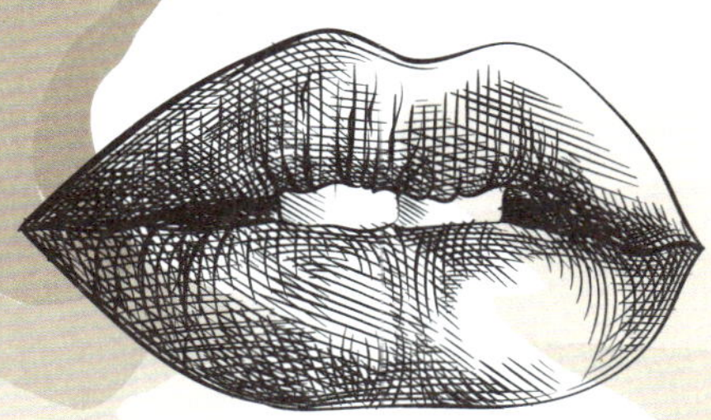

[3과제] 캐릭터 메이크업

시험시간 50분

Point

레오파드 : 한 톤 밝은 베이스
엘로우, 오렌지, 브라운 컬러로 그라데이션
검은색의 레오파드 무늬
버건디 레드의 인커브의 립

한국 무용 : 매트한 피부
부드러운 곡선의 아이 브로우
마젠터 컬러의 상승형 아이 쉐도우
검은색 귀밑머리

발레 : 매트한 피부
갈마기 눈썹
아쿠아 블루, 화이트로 아이쉐도우
언더 라인과 눈과빈공간 화이트

노역 : 한 톤 어두운 매트한 피부
갈색 주름
회갈색 아이 브로우
네추럴 베이지 립

Make-up Title : 한국무용		Skin Color : 크림 화운데이션을 이용하여 완벽한 커버	
Eye-Shadow			
High-Light Color	Main Color	Point Color	Accent Color
White	Pink	Magenta	Black
Cheek Color		Lip Colcr	
Pink		Red	

Make-up Title : 발레		Skin Color : 피부톤(크림 화운데이션사용)	
Eye-Shadow			
High-Light Color	Main Color	Point Color	Accent Color
White	Pink	Purple	Aqua Blue
Cheek Color		Lip Color	
Pink		Rose Pink	

Make-up Title : 레오파트		Skin Color : 밝게 표현(크림 화운데이션사용)	
Eye-Shadow			
High-Light Color	Main Color	Point Color	Accent Color
White	Orange	Brown	Black
Cheek Color		Lip Color	
		Dark Red	

Make-up Title : 노역		Skin Color : 밝피부톤 약간 어둡고(matte)	
Eye-Shadow			
High-Light Color	Main Color	Point Color	Accent Color
Light Beige	Brown		
Cheek Color		Lip Color	
		Natural Beige	

[4과제] 속눈썹 익스테션 및 수염

시험시간 25분

Point

수염

- 마네킹에 스프리트 껌 좌우 대칭으로 바를 것
- 아래턱선부터 수염을 붙임
- 양 볼 쪽으로 수염양을 줄여가며 붙임
- 콧수염 양쪽 끝부터 붙이며 중앙으로 기울기 세우며 붙일 것.
- 수염의 길이 1~2cm정도
- 스프레이로 수염형태 고정

Make-up Title :		Skin Color :	
Eye-Shadow			
High-Light Color	Main Color	Point Color	Accent Color
Cheek Color		Lip Color	

Chapter 17

Make up Illustrations

Chapter 18

Entertainer Make up Style

2018.09.18

T 2018.09.17

J 2018. 09. 14